Astronomers' Universe

More information about this series at
http://www.springer.com/series/6960

Jerry L. Cranford

Astrobiological Neurosystems

Rise and Fall of Intelligent Life Forms in the Universe

 Springer

Jerry L. Cranford
Department of Communication Disorders
LSU Health Sciences Center
New Orleans
Louisiana
USA

ISSN 1614-659X   ISSN 2197-6651 (electronic)
ISBN 978-3-319-10418-8   ISBN 978-3-319-10419-5 (eBook)
DOI 10.1007/978-3-319-10419-5
Springer Cham Heidelberg New York Dordrecht London

Library of Congress Control Number: 2014950889

Cover illustration: The Orion Nebula, the closest of over 100 known huge gas and dust clouds located in our Milky Way Galaxy, appears to be a very busy "newborn nursery" where many new solar systems are being built. This photograph courtesy of NASA/ESA/Luca Ricci shows a collection of six such new planetary systems, some of which may in the far distant future, provide a home for life. Credit: NASA, ESA, M. Robberto (Space Telescope Science Institute/ESA), the Hubble Space Telescope Orion Treasury Project Team and L. Ricci (ESO)

Printed on acid-free paper

Springer is part of Springer Science+Business Media (www.springer.com)

Preface

Until close to the end of the twentieth century, man had no evidence that living creatures of any kind could exist anywhere else in the universe other than our own planet. Due largely to the advent of computers and rocket science, in the last years of the twentieth century, scientists began making a series of amazing discoveries that have completely revolutionized the biological sciences as well as the geological and space sciences. Many scientists now believe we are only a few years away from confirming the existence of extraterrestrial life not only in our solar system but also possibly on planets circling other stars in our galaxy, and even beyond our galaxy.

Upon retiring in 2008 from a 40-year career as a professor of psychology and the brain sciences, the author immediately elevated what had been his lifelong hobby as an amateur astronomer from a part time to a fulltime passion and is now devoting his retirement years to pursuing a second career (albeit without pay) as an amateur astrobiologist. In 2011, I published my first book on this topic entitled *From Dying Stars to the Birth of Life: The New Science of Astrobiology and the Search for Life in the Universe* (Nottingham University Press, United Kingdom) for interested nonscience readers. While writing the book, which was designed to be an overview of the new science of astrobiology, with a particular emphasis on how life and Earth's geology coevolved to produce a life friendly world, I frequently found my "alternate self" (i.e., my brain science/psychology background) tricking me into speculating about how different kinds of intelligent life might have arisen if Earth's earlier environmental conditions had been different in some way. I decided that my next book should take advantage of my knowledge of both the space and brain sciences to see if I could come up with any new ideas about how intelligent life might be able to evolve on other worlds.

In the present book, I have taken the bold step of attempting to merge my knowledge of astronomy and the neurosciences and describe my personal (plus professional) thoughts on how intelligent nervous systems might be able to develop and survive on other planets (or moons) in the universe. For many years, those astronomers who were interested in investigating the possibility of life on alien worlds (who were first labeled as *exobiologists* but now call themselves *astrobiologists*) were accused of pursuing a science that was totally devoid of any observable or measurable subject matter which they could directly investigate (no extraterrestrial life-forms, no exoplanets, etc.). When I first started writing the present book in January, 2012, I realized that I would be faced with the same predicament that haunted the astrobiologists since I also had no hardcore data to directly support any ideas, theories, or speculations I might want to discuss concerning the possible existence of intelligent extraterrestrials. However, the good news is that our astrobiologists and space scientists now appear to be very close to putting an end to this longstanding dilemma. While we still have no confirmed evidence for the existence of life of any kind, including intelligent life, on any other planet other than Earth, our astrobiologists are now finding solid evidence that potential homes for life (i.e., exoplanets) may be relatively common in the universe, and it now appears likely that it is only a matter of time until we discover Earth-like worlds circling stars in life friendly warm and wet habitable zones that could support carbon-based life similar to that which inhabits our planet.

Therefore, while I have knowledge (and access) to a considerable amount of scientific data related to how life and intelligent nervous systems evolved on Earth, I know nothing about how (or even whether) such systems might evolve on other worlds. This book will, therefore, be mostly filled with my educated guesses, as well as those of other respected Earth-bound scientists, of what may be happening out there beyond our atmosphere. When there is little or no confirmed data to support a field of science, it is disturbingly easy to be wrong in describing how things might happen. Some rare individuals sometimes manage to avoid this pitfall as did Charles Darwin whose ideas on biological evolution were amazingly on-target in spite of the fact that he knew virtually nothing about the underlying chemical, genetic, and biological underpinnings of this field of science. I am hoping that I might be lucky enough to get close to the truth with some of the many speculations that this book will, of necessity, contain.

Another major reason I chose to write this book at this time is that I believe very strongly that, even though we have not yet confirmed the existence of extraterrestrial life, It is not too soon to take the next major step in unraveling the mysteries of this exciting new field of science by inviting our behavioral scientists (neuroscientists, psychologists, and even sociologists) to join ranks with our astrobiologists to begin serious discussions and debate on what kinds of intelligent life-forms may exist out there and how we might go about locating and studying them. The amazing invention of digital computers in the last half of the twentieth century not only opened the door wide open to a bona fide scientific revolution in my childhood hobby of astronomy (and space science), it launched a similar revolution in the waning years of my professional career as a neuroscientist/psychologist. The recent advent of fast and powerful computers is also now triggering an incredible explosion of fascinating new information on how the brain works. As a newly retired professor, I find myself literally caught between the proverbial "rock and a hard place" and have decided to devote my "golden years" to learning as much as I can from my younger and still active older colleagues (plus, hopefully, my readers) about whether brains on other planets may work the same or entirely differently from those found on Earth.

Once again, as I did in my last book, I cannot take personal credit for any of the data, theories, ideas, etc. that this book contains with respect to astronomy and astrobiology. And, although I was an active brain researcher and published numerous research articles and a few textbooks, the brain science information presented in the present book is also public domain. Finally, I would again like to thank my wife Fran for tolerating the way I have converted what was supposed to be our leisure retirement years into a busy second career.

New Orleans, LA   Jerry L. Cranford

Contents

Chapter 1
Scientists Believe Intelligent Life May Be More Common in the Universe than Previously Considered Possible

The construction of giant telescopes at the beginning of the twentieth century combined with the advent of digital computers and rocket science in the last part of the twentieth century totally changed mankind's thoughts about how common life, and especially intelligent life, may be in the universe. Our knowledge of the physical size of our universe suddenly exploded in 1925 when the astronomer Edwin Hubble looked through what was then the largest and most powerful telescope (Mt. Wilson Observatory in California) in the world and discovered the existence of galaxies located outside our own Milky Way galaxy (Fig. 1.1a, b). Up to that point in time, most astronomers believed our Milky Way galaxy was itself the whole universe, with nothing existing beyond the most distant stars we could see with our best telescopes.

Astronomers today believe that our Milky Way galaxy is but one of a huge number of other galaxies that fill a universe that is incredibly large. While few scientists have been so bold as to suggest that the universe may be infinite, a few have actually posed the idea that other independent universes may exist (Kaku 2006; Greene 2011; Gribbins 2009). Many astronomers, however, now estimate that there are possibly as many as 200,000,000,000 (two hundred billion) or more other galaxies out there, some of which are smaller and others which are larger than the Milky Way galaxy (Bennett et al. 2003; Bennett and Shostak 2011; Chaisson and McMillan 2000). And, believe it or not, our own average size Milky Way galaxy is so large that it takes light 100,000 years (traveling at a speed of 186,000 miles/s) to travel from one side of the galaxy to the other side. And our astronomers now believe that many of those other galaxies in the universe may contain as many as 200 billion or more stars and, once again, some of their stars are smaller than our own sun, while others are tens, hundreds, or even thousands of

© Springer International Publishing Switzerland 2015
J.L. Cranford, *Astrobiological Neurosystems*, Astronomers' Universe,
DOI 10.1007/978-3-319-10419-5_1

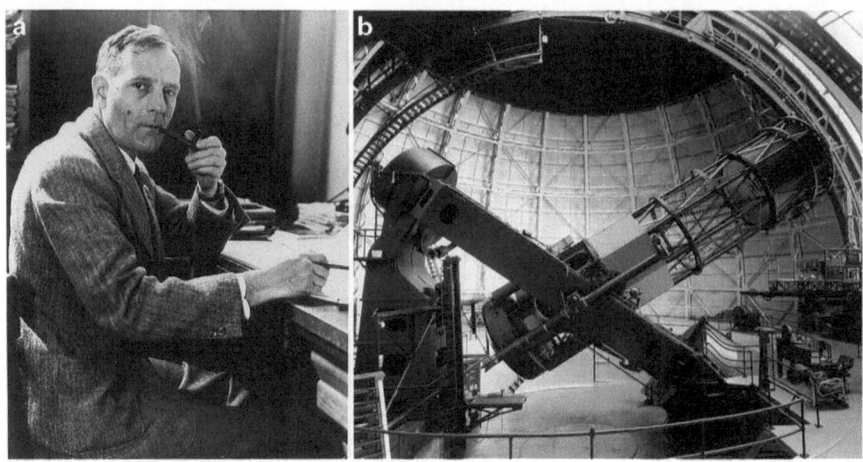

Fig. 1.1 (**a**) In the early 1920s, the astronomer Edwin Hubble used the world's largest telescope located at (**b**) the California Institute of Technology's Mount Wilson Observatory to discover that the universe consists of large numbers of more distant galaxies other than our Milky Way galaxy. Edwin Hubble (1889–1953) and Hooker Telescope (2.5 m), Mt. Wilson Observatory. Sources: Wikipedia, http://www.astro.caltech.edu/ (image credits: Wikipedia Commons/Caltech/Huntington Library)

times larger.[1] And, when our astronomers use the world's most powerful space telescope (the Hubble telescope named after the astronomer Edwin Hubble) to look out into the distant regions of space, they are amazed that the light from the most distant galaxy they can see (which, through the telescope, looks like a tiny dim speck of light) required about 13.2 billion years (again, of course, traveling at a speed of 186,000 miles/s) to get to the Hubble telescope. And, even more amazing is the fact that many of our best scientists now tell us that this tiny speck of light does not necessarily mean we are looking at the far distant edge of our universe, but that the light from other more distant galaxies or celestial objects has not yet had enough time to reach us! In recent years, a few astronomers and theoretical physicists (cosmologists) have even gone so far as to propose that, instead of just one single universe that was created and started expanding 13.7 billion years ago, the total universe itself may consist of many multiple universes (i.e., "multiverses") or even other so-called "parallel" universes (Vilenkin 2006) that are located far beyond the small part that we can see today with our best telescopes.[2] If they do

---

[1] The largest known star in our Milky Way galaxy is VY Canis Majoris, which is located 5,000 light years from the Earth and has a diameter that is estimated to be approximately 2,000 times greater than our own sun. And, at the other end of the size scale, there are many stars that are less than 1/10 the diameter of our sun which are known as "brown" dwarfs since they are not large and massive enough to trigger nuclear fusion in their cores and, therefore, are not very bright.

[2] Having said this, I feel I must also add that some very competent scientists also believe that the leftover remnants of some of the largest exploding stars, i.e., those invisible things we call "black holes", may be the gateways to these other multiverses or parallel universes. Some scientists have even been bold enough to suggest that some of these other universes may operate using completely

exist, it may require many trillions of years before the light from these more distant objects would have time to reach our telescopes. And, since our sun, as part of its normal evolution, is continuing to get hotter and hotter with time, we only have about another billion years left before it begins to slowly become too hot to sustain life. If mankind wants to survive long enough to be able to see these more distant parts of our vast universe, we will need to relocate our species (migrate) to cooler locations elsewhere in our universe in a few billion years and patiently wait for the light from these objects to get to us.

If the universe is so unbelievably large, how can astronomers possibly measure distances from one point in space to another? Because our universe is so huge, astronomers cannot easily measure distance in kilometers or miles. Although our closest neighbor in space, our own moon, is only 238,900 miles away, the distances to the next closest objects (planets) in our own solar system requires us to talk in terms of millions or even billions of miles. And, shortly after 1900, it got totally insane when we started measuring distances to other stars in our own Milky Way galaxy. The closest star to us, other than our sun, is Proximi Centauri, which is only **25,689,592,881,951** miles away (i.e., **25.7 trillion** miles).[3] And, after Edwin Hubble and other astronomers informed us in the 1920s that our Milky Way galaxy is not the only galaxy in the universe, things got even crazier. Some astronomers now believe that, of the 200 billion or more other galaxies we think are out there, the **most distant galaxy** we can see (i.e. that distant tiny speck of light that the Hubble telescope found) may be a staggering **767,656,960,000,000,000,000,000** miles away! So, astronomers suddenly needed to come up with a new way of measuring such extreme distances. Since light is the fastest known thing in the universe and is fast enough to actually travel around (circle or orbit) our world **seven times in 1 s**, the astronomers chose the total distance light can travel in one calendar year (365 days) as their new unit of measurement. So, using this concept of *light years*, the distance to Proximi Centauri suddenly became 4.3 light years away, and the most distant galaxy we can see with our best telescopes suddenly became **13.2 billion** light years away.[4]

Now that the reader has some idea of how large our universe may be, it is time to try to explain how old our universe may be and "how it got here"! This is the topic area in our science of astronomy that even our best scientists admit they know the least about and is the most puzzling. Many astronomers currently believe that the whole universe that we can see today with our best telescopes was created about

---

different laws of physics and chemistry from those of our own universe and some may even be totally devoid of this thing we earthlings loosely refer to as life.

[3] In my last book (*From dying stars to the birth of life*) I somewhat jokingly, but quite seriously, indicated that traveling to Proximi Centauri in a modern Jumbo jet, if such a thing was possible, at an average speed of 600 miles/h would require 4.6 million years to achieve.

[4] Thus, when our astronomers look at Proximi Centauri through their telescopes, they are seeing this star exactly as it appeared 4.3 years ago, and that tiny speck of light that is 13.2 billion light years away is what that galaxy looked like long before humans or even the planet we live on was even created.

13.7 billion years ago in what scientists call a **Big Bang** event (Delsemme 1998). Almost all scientists will admit that they are almost totally clueless as to what, if anything, might have been present before this Big Bang thing happened. Of the basic components of the universe that we know about, i.e., *matter, energy, space*, and even *time*, only energy is believed to have been present before the Big Bang, and it is believed to have existed as an infinitely small, dense, and hot "glob" or "speck" of some kind of pure energy. Many scientists today believe this small precursor to our universe was actually much smaller than a single atom! The Big Bang event was not an explosion in the traditional sense of the word but some kind of sudden and rapid expansion of this pre-existing incredibly small and dense "piece" of energy into everything that we can see in the universe today.[5] The universe is continuing to expand in size even today with the more distant galaxies still racing away from us. While the idea that our known universe was created about 13.7 billion years ago in a gigantic expansion (but definitely not an explosion) or what some scientists call an "inflation" from an unbelievably small piece of nothing but pure energy seems almost too bizarre for the author and many other scientists to believe, all of our best scientific tools have been repeatedly and persistently telling us for the past 60+ years that it really did happen! Reality is indeed sometimes stranger than fiction.

Thus, the growth of mankind's knowledge of astronomy and the universe we live in between the early years of the twentieth century and the beginning of the new twenty-first century has been truly "astronomical" in every sense of the word (Bennett et al. 2003; Chaisson and McMillan 2000). When I was a student in elementary school in the 1950s, my science teachers told me that life was both complex and fragile and outer space was so hostile to all living things that mankind might be the only intelligent life in the entire universe. And, for entirely different reasons, my Sunday school teacher seconded this opinion. Now, at the beginning of the new twenty-first century, our scientists are suddenly telling us that life may be tougher, plus more flexible and resilient, than we would have dared imagine possible just a few short years ago. And our astronomers have suddenly started discovering that planetary homes for extraterrestrial or alien life may be common throughout the universe (Aguilar 2013; Bortz 2008; Darling 2001; Gilmour and Sephton 2003; Lunine 2004; Plaxco and Gross 2006). In this chapter I will describe these exciting new discoveries of our life and space scientists that have so drastically changed man's beliefs about life in our universe?

---

[5] Some scientists, including Steven Hawking, even believe that, in addition to space, energy, and matter, time itself also did not exist before the Big Bang. If this is so, then in a very real sense **NOTHING** existed before the Big Bang! Dr. Hawking believes that, since nothing existed before the Big Bang, including time, the universe could not have been created by a Higher Power.

Stars and Planets Are Born Together in Collapse of Large Interstellar Gas and Dust Clouds

For thousands of years, mankind believed the Earth was the center of the universe, and all the objects (sun, moon, stars) we could see in the sky revolved around us. While a Polish astronomer named Nicholas Copernicus managed over 500 years ago to "demote" the Earth as being the center of the universe, much of the world's population continued to believe Earth might still be totally unique in being the only place life could exist. In 1916, an astronomer named James Jean proposed a theory of solar system formation (known as the "Tidal Theory") that endorsed the idea that our solar system was the result of an extremely close encounter between our sun and another star that happened to wander into our neighborhood. While the other star did not collide head on with our sun it came close enough that the two stars glanced off (grazed) each other which resulted in some of the material from our sun being torn away to form a cloud of debris that began to orbit around (i.e., circle) the sun and eventually condense into a family of planets. Since the average distance between any two stars in our region of the Milky Way galaxy is extremely large, the probability of any two stars colliding is virtually zero. Therefore, if this earlier theory of solar system formation had turned out to be accurate, the possibility of other Earth-like planets that might allow life would be very improbable. However, by 1940 this "close encounter" theory of planetary formation was finally relegated to history's "outbox" (recycle bin). Most astronomers now believe that stars and planetary systems are formed together when a supernova explosion (or other kinds of interstellar disturbance) produces a shock wave that causes any nearby interstellar gas and dust clouds to collapse and form much smaller spinning protoplanetary disks that eventually condense into new stars and planetary systems.(Cranford 2011; Casoli and Encrenaz 2007). Figure 1.2 shows how many astronomers now believe at least one form of this collapsing planetary disk formation process works. Although not all stars are believed to develop planetary systems, scientists now estimate that as many as 50 % of all sun-like stars may develop such systems (Irwin 2008; Jayawardhana 2011; Mason 2008).

Although our scientists had long been well aware that our universe contains huge numbers of other galaxies each containing incredible numbers of stars, this new idea that planetary systems might be common did not catch on real fast. As late as 1995 mankind's egocentric nature still convinced many of us that our sun might be, if not the only one, at least one of the very few stars which have planets circling it. In that year, astronomers Michel Mayor and Didier Queloz at the University of Geneva in Switzerland spotted a star in their telescope that appeared to have a very small wobble in its motion through space (Fig. 1.3). While they could not see it, the scientists measuring instruments indicated that this small wobble was caused by a planet that was gravitationally tugging on the star as it circled it. Mankind suddenly realized that ours is not the only planetary system in the universe—there was now at least one other one out there. This finding by Swiss astronomers of what is now

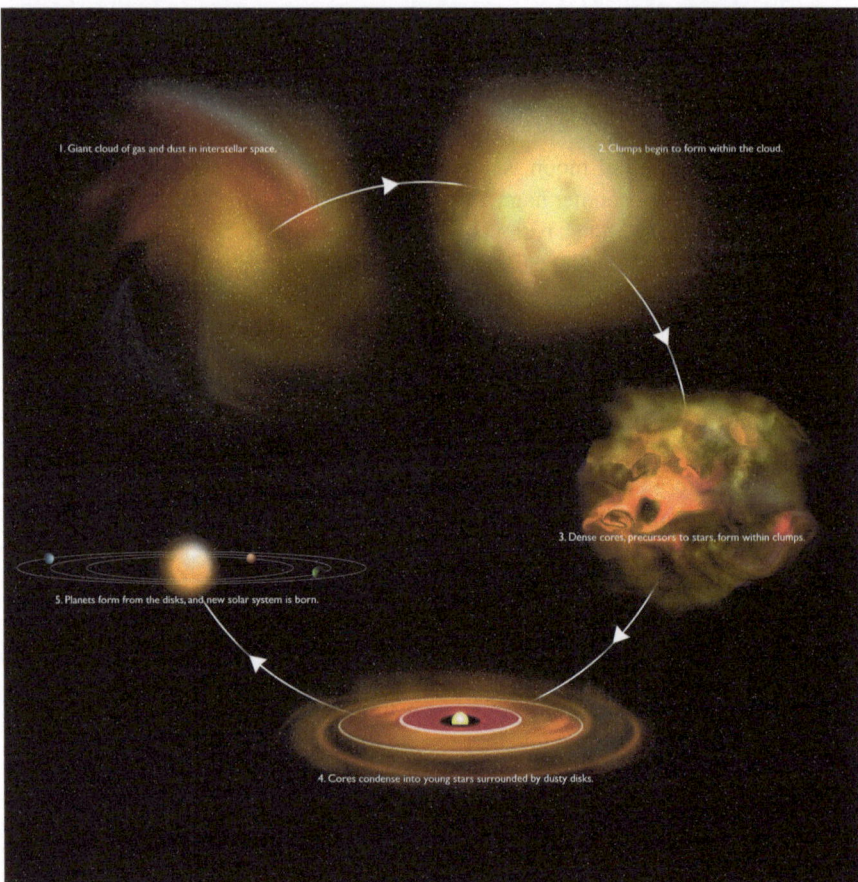

Fig. 1.2 Shows an artist's drawing of how astronomers believe stellar systems (stars and orbiting exoplanets) develop from giant interstellar gas and dust clouds. Following a nearby supernova event, the giant gas/dust cloud breaks up into smaller individual clouds which then begin to gravitationally collapse in on themselves and, at the same time, begin to spin. As the smaller cloud contracts, it starts spinning faster and begins to take on the form or shape of a flat pancake with a bulge in the center (the location of the new proto-star). As the accretion disc spins even faster, a series of concentric rings of dusk clouds begin to form. These rings will eventually become new planets. The dust in the spinning rings begin to collide and stick together and grow into increasingly larger particles, then planetesimals ("miniature planets"), and finally full-sized planets. After the planets are formed, most of the leftover gas, dust, and debris are removed from the new planetary system (image credit: Bill Saxton, artist, National Radio Astronomy Observatory)

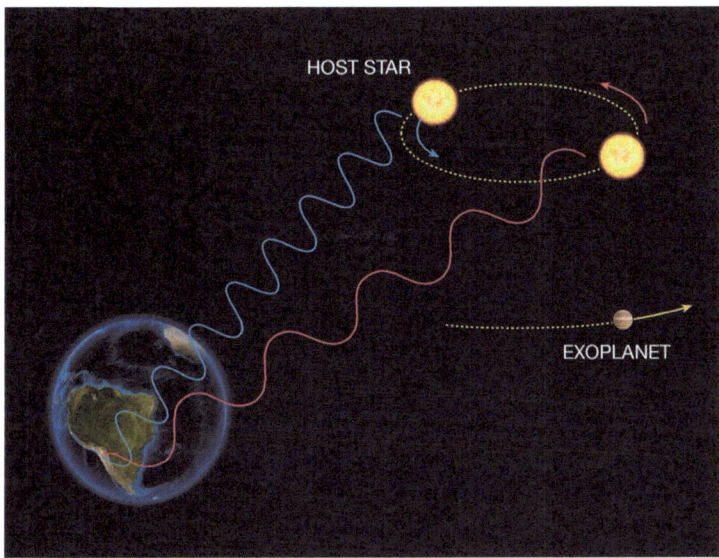

Fig. 1.3 Illustrates how the Doppler affect can be used to detect the presence of exoplanets orbiting stars. The exoplanet produces an extremely small gravitational tugging effect on the star as it orbits it which causes it to "wobble" or deviate from a straight line in its movement through space. When the star wobbles in the direction of the Earth, the light waves from the star get closer together and become shorter in wavelength and the star becomes slightly more blue in color. When the star wobbles away from Earth the light waves get further apart and the star's light becomes redder (image credit: European Southern Observatory)

known to be the third exoplanet discovered to be orbiting another star[6] triggered a huge wave of excitement among the world's astronomers. Astronomers started looking for more exoplanets, *and more exoplanets they quickly found!* By the year 2000 astronomers had discovered 40 other such worlds, and by 2010, this number had grown to over 500, with some stars having families of planets similar to our own solar system (Irwin 2008; Mason 2008; McKay 2008). At first, the only planets that could be found were giant gas planets (like our own Jupiter or larger) simply because they produced larger tugging effects on their home stars that were more detectable. As these new ***Planet hunters*** (as they were now whimsically

---

[6] The first actual exoplanets were discovered by radio astronomers in 1992. Two large (Jupiter size) planets were found to be orbiting a small extremely massive type of star known as a "pulsar". Pulsars are the left over remnants of giant stars that have undergone supernova explosions. Such stars are extremely small (only a few miles in diameter) and extremely massive (weighing as much as the original star before the supernova). They also rotate (spin) at incredibly fast rates and emit huge amounts of deadly radiation that would fry any life on any nearby planets. Whether exoplanets circling pulsars are formed before the occurrence of supernova events or afterwards is unknown. It is ironical, however, that the first exoplanets discovered by astronomers are, by any scientific criteria we know of, totally hostile to any known forms of life our best life scientists currently believe could exist. Subsequent exoplanet discoveries have continued to find that the majority of exoplanets are also unfriendly, at least by mankind's current standards. While life may not be rare in the universe, human-like critters may be quite rare.

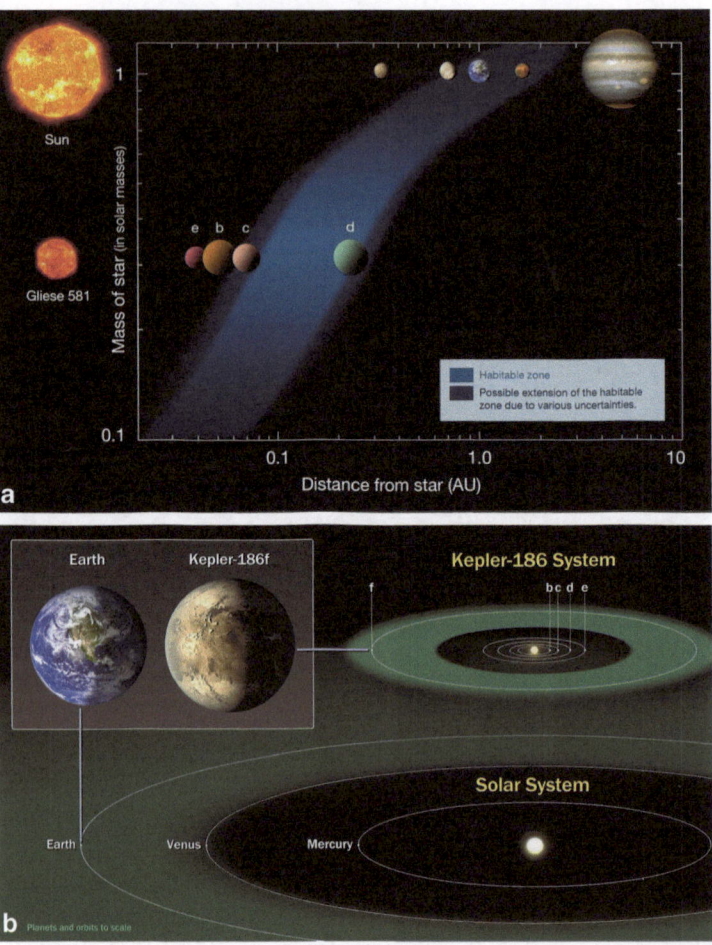

Fig. 1.4 (a) Depicts the location of the habitable zones for our own solar system in contrast to that of the Gliese 581 exoplanetary system (image credit: European Southern Observatory), while (b) *shows a NASA artist's drawing of the first "twin Earth" (that is only 10 % larger than Earth) which was discovered by astronomers on April 18, 2014! This planet orbits a star in its habitable zone, which would be warm enough for our type of life, if it were to have an atmosphere similar to ours. Unfortunately, any visitors from this planet would have to travel at the speed of light for 490 years to visit us! The exoplanet orbits a small red dwarf star (slightly smaller than our sun) which would make it ideal as a habitable world* (image credit: NASA Ames/SETI Institute/JPL-Caltech)

labeled by the news media) improved their detection techniques they rapidly began finding smaller and smaller exoplanets that produced even smaller but still detectable tugging effects. In the last few years the planet hunters have started finding "*super-Earths*" (Fig. 1.4a) that, in some cases, are only about two or three times larger than Earth, with some orbiting in their home star's "habitable zones" where liquid water and life friendly surface conditions may exist (Sasslov 2012). And, just recently, scientists identified the first possible "twin" of our Earth, which is almost identical in size to our home planet plus also orbiting in the middle of its home star's habitable or "goldilocks" zones (Fig. 1.4b). The habitable zone for any given planet, as defined in astronomy and astrobiology, is the cooler (warm but not hot) region around a star within which it is theoretically possible for planets with sufficient atmospheric pressure to maintain liquid water on their surfaces. Since liquid water is essential for all known forms of life, planets in this zone are considered the most promising sites to host extraterrestrial life. The habitable zone for any ETs that are not dependent on the carbon atom and water would require a totally different definition. In recent years, growing numbers of our life and space scientists have begun to believe that such "extremophile" type ETs may be out there!

Astronomers Suddenly Finding Potential Planetary Homes for Alien Life May Be Common in Universe

Until the spring of 2009, the biggest obstacle to finding smaller Earth-like planets circling other stars was that all of our telescopes were Earth-bound which forced us to deal with a dirty and turbulent atmosphere that made it difficult to see the really small stellar wobbles produced by the smaller exoplanets. However, in spite of this limitation, by that time astronomers had already been able to use land-based telescopes to identify close to 500 actual exoplanets circling other stars in our galaxy. Starting in 2007, astronomers began using the powerful new Keck Observatory twin telescopes in Hawaii to search for exoplanets. The Keck telescopes were using recently developed high speed computer adaptive optical technologies that allowed them to cancel out much of the atmospheric interference that had long plagued ground-based telescopic observations (Fig. 1.5a, b). In 2007, the Keck astronomers discovered a red dwarf star (Gliese 581) located only 20 light years away from Earth in the constellation Libra which is now thought to possibly host as many as six orbiting exoplanets. This amazing discovery caused many of our astronomers to quickly trade in their astronomer hats for planet hunter hats when it became apparent that three of these exoplanets were rocky planets close in size to Earth that appeared to be orbiting in their home star's habitability or goldilocks zone which suggested they might be warm and wet enough to support some kind of carbon-based life. The discovery of this possibly life friendly stellar system that was virtually a next door neighbor in our galaxy not only excited the astronomy world but triggered a flood of television and popular news media reports that began pushing the idea that man might not be alone in the universe.

On June 23, 2013, as the present book was in the final stages of being prepared for publication, and the scientific community was in the throes of excitement from the incredible numbers of new exoplanets suddenly being discovered by NASA's new Kepler space telescope (NASA 2013), astronomers learned that land-based telescopes were still quite capable of discovering even more exoplanets. The same group of astronomers that had earlier reported evidence that Gliese 581 may host

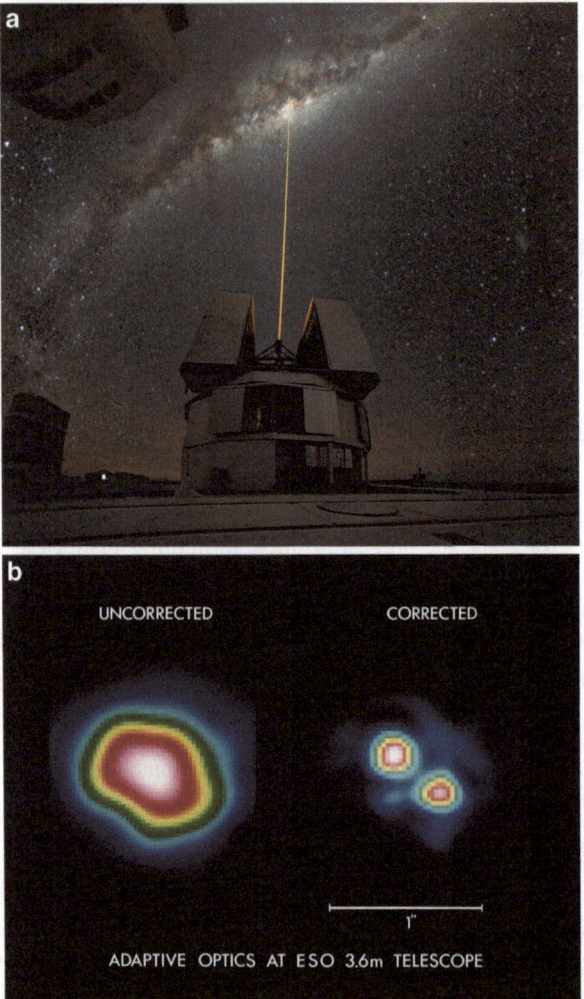

Fig. 1.5 Thanks to high speed computers and lasers, astronomers can now reduce the distortion in their telescopic images that result from thermal induced movement (shimmering) of air molecules. The adaptive optical technique involves (**a**) pointing a steady laser beam at a star located close to where the telescope is aimed. When the star twinkles due to thermal distortion, the laser beam will continually monitor (measure) the amount of visual distortion in the star's image and transmit this information to a computer which will immediately send orders to the telescope mirror and "tell it to distort itself" appropriately to compensate for the star's distortion. Today's largest observatory telescopes have mirrors that are made up of very large numbers of separate smaller mirrors that are positioned adjacent to each other but can be made to slightly change their orientation angle (i.e., "distort" themselves) relative to neighboring mirrors to quickly compensate for the distortion that the laser beam has detected in the star's light (image credit: European Southern Observatory, Wikipedia Commons). Finally, (**b**) shows the amount by which stellar images can be cleaned up by this laser-based adaptive technique. The image on the *left* was taken with the adaptive optical tool turned off, while the image on the *right* was taken with the adaptive optical system turned on. The large fuzzy blob on the *left* turned out, as shown on the *right*, to be a double star system. Adaptive optics along with the use of multiple telescopes (i.e., *interferometry*, which I will describe in Chap. 2) is keeping our land-based telescopes in the planet hunting game! (image credit: ESO)

two or three life friendly exoplanets, reported they had now identified a triple star system that is located about the same distance from Earth as Gliese 581 (at a distance of 22 light years rather than Gliese 581's distance of 20 light years) that may also have super-Earth type exoplanets orbiting it.[7] These astronomers reported that they had combined astronomical data gathered from the Keck telescopes with that collected using the world's largest land-based interferometry telescope system, the new European Southern Observatory's (ESO) Very Large Telescope Array (VLT) of four telescopes in southern Chile and had discovered the existence of a group of five exoplanets that are circling the smallest member (Gliese 667C) of a triple star system that is located in the Scorpius constellation.[8] Three of these latest exoplanets that are circling Gliese 667C are slightly larger than Earth (super-Earths) and have been confirmed to be orbiting in their home star's habitable zone which might allow the presence of life friendly surface conditions (Fig. 1.6a). The other two exoplanets that are orbiting Gliese 667C (Gliese 667Cb and Gliese 667Cd may be too hot or too cold respectively to support carbon and water based life (Fig. 1.6b). Finally, Fig. 1.6c shows an artist's sketch of what Gliese 667Cc might look like close up with the other two stars of the Gliese 667 triple star system shown in the upper left corner of the drawing. These new findings with the Gliese 667 triple star system now suggest, in combination with the finding of possible life friendly planets circling Gliese 581, that twin Earths may be more common in the universe than many astronomers previously believed.

On March 6 of 2009, the identification of exoplanets suddenly became much easier when the National Aeronautics and Space Administration (NASA) launched the first space telescope (Fig. 1.7a, b) that would be totally dedicated to detecting exoplanets (NASA 2013). This **Kepler mission** (named after a famous seventeenth century mathematician and astronomer) started using a second method for detecting exoplanets that was even more sensitive than searching for wobbling stars.

---

[7] However, for retired professors like me whose memory skills may have declined slightly, I need to alert my readers that, in addition to boggling our brains with talk of extreme size, extreme distance, extreme numbers, and even extreme periods of time (i.e., "deep time"); the new planet hunters have now come up with one more means of confusing us. When they label a single star they typically tell us which star catalog it is listed in (e.g., the Gliese catalog) and give it a number, e.g., Gliese 667. When the star is a member of a double or triple star system, they then use capital letters to distinguish which member of the system the star is, e.g., Gliese 557A or Gliese 667B, if a double star, and Gliese 667C if it is the third member of a triple star system. When a single star has one or more exoplanets orbiting it, the astronomers use lower cap lettering to designate which one it is, e.g., Gliese 581a or Gliese 581b or Gliese 581a, b, c, etc. if they want to list more than one of the planets that are circling the star. Now, if a member of a double or triple star has planets orbiting it, the astronomers then label it as Gliese 667Aa, or Gliese 667Ab, and so on, to designate which star in the star system is being referred to as well as which exoplanet is being referenced. If I ever see a reference to some star like Whoopi 101Cz, I will definitely assume life is probably rampant in the universe!

[8] The reason these two star systems are named "*Gliese 581* and *667*", respectively, even though located in different parts of the night sky, is that they are listed in the same *astronomy star catalog*, that was published by Wilhelm Gliese, a famous nineteenth century German astronomer. This particular star catalog only lists stars that are located at a distance of 65 light years or less.

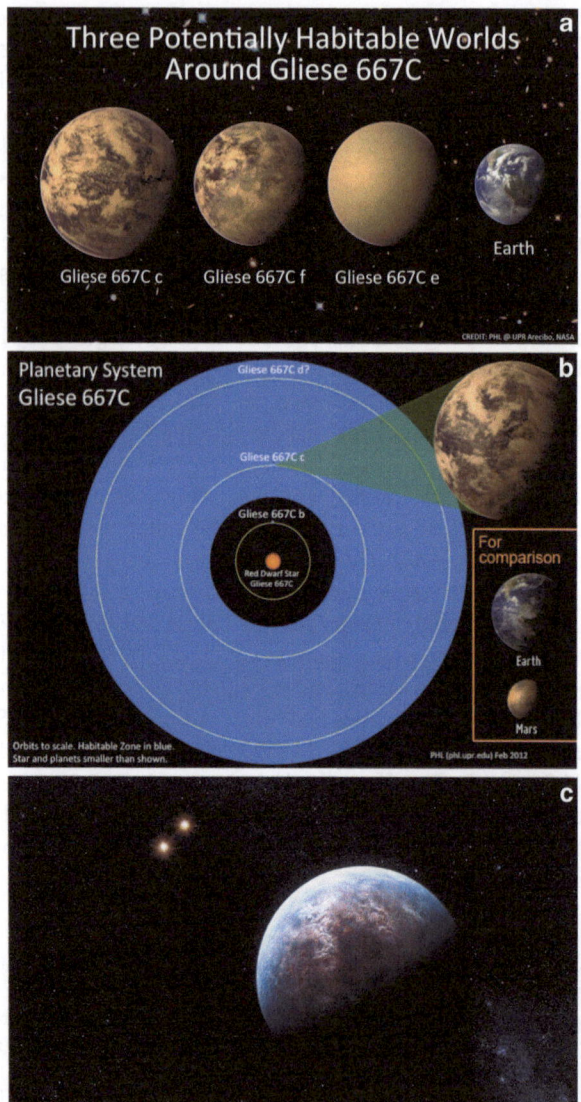

Fig. 1.6 It seems that, in spite of all the excitement over the use of space telescopes, land-based telescopes are not yet ready for "early retirement" when it comes to hunting for exoplanets. Gliese 667 is a triple star system with three stars (Gliese A, Gliese B, and Gliese C). Of the three stars, Gliese C is the smallest and is a red dwarf star that appears to have five exoplanets orbiting it. (**a**) shows an artist's drawing of the three super-Earth sized exoplanets that were just recently discovered using land-based telescopes (the European Southern Observatory and the Hawaiian Keck telescopes) to be orbiting in the life friendly habitable zone of Gliese 667C (image credit: PHL @ Arecibo, NASA). (**b**) Shows a schematic drawing of the location of the orbit of potentially habitable Gliese Cc along with those of Gliese Cb and Gliese Cd which may be too hot or cold to harbor life (image credit: Planetary Habitability Laboratory @ UPR, Arecebo). While (**c**) shows an artist's depiction of what Gliese 667Cc might look like close up with the other two stars of the Gliese 667 triple star system (i.e., Gliese 667A and 667B) shown in the *upper left part* of the drawing

Fig. 1.7 (**a**) Shows a photograph of the Kepler Space Telescope being launched on March 6, 2009 by a NASA Delta rocket in Florida, while (**b**) shows a NASA artist's drawing of the Kepler space telescope as it begins to search for transiting exoplanets in the intended search area shown on the *right* side of the image. The formal search area for the Kepler telescope is a very small region of space that approximates only 1/400th of the total area of the entire sky that could potentially be viewed from Earth (including both the north and south hemispheres) This small area of the night sky was selected because it is located in the most "crowded part" of the Milky Way galaxy which has a higher concentration of stars than anywhere else in the galaxy (image credits: NASA)

Astronomers now began looking for extremely small reductions in the total light coming from a star that occurs when an exoplanet moves in front of it and blocks some of its light making it appear slightly dimmer (Fig. 1.8a, b). This effect is similar (but much smaller) to what happens when the moon moves in front of and "eclipses" our sun. Because astronomers now did not have to deal with the interference of the atmosphere they could actually see reductions in a star's light as small as 1/100 of 1 % that would indicate the presence of exoplanets as small or smaller than our Earth.[9]

The Kepler telescope, in just four short years from 2009 to 2014, has now made the "explosion" of exoplanet discoveries that was begun in 1995 by ground-based telescopes look like the wimpy "pop" of a small firecracker! As of November 2011, using both the new Kepler space telescope and our best land-based telescope systems, the planet hunters reported that their total count of possible candidates for exoplanets had exceeded 1,025 with 16 of these considered to possibly be life friendly super-Earths. And two short years later, On June 26, 2013, the Kepler astronomers released a second list that now contained 12 more possibly habitable exoplanets (Fig. 1.9). Yet, since NASA has Kepler only looking at one very small sector of the sky that is only about 1/400 of the total sky area that is visible from Earth (Fig. 1.10), it is quite likely that the 2,000+ exoplanets so far detected by

---

[9] On June 5, 2012, the planet Venus moved in front of the sun (transit event) and produced a reduction in the total brightness of the sun of approximately 0.1 % (one in a thousand parts, i.e., 1/1,000). While, with the appropriate equipment to protect the eyes (special sun blocking lenses) from the sun's harmful rays, it is possible to see Venus crossing in front of the sun, the human eye is not capable of detecting a brightness change that is this small.

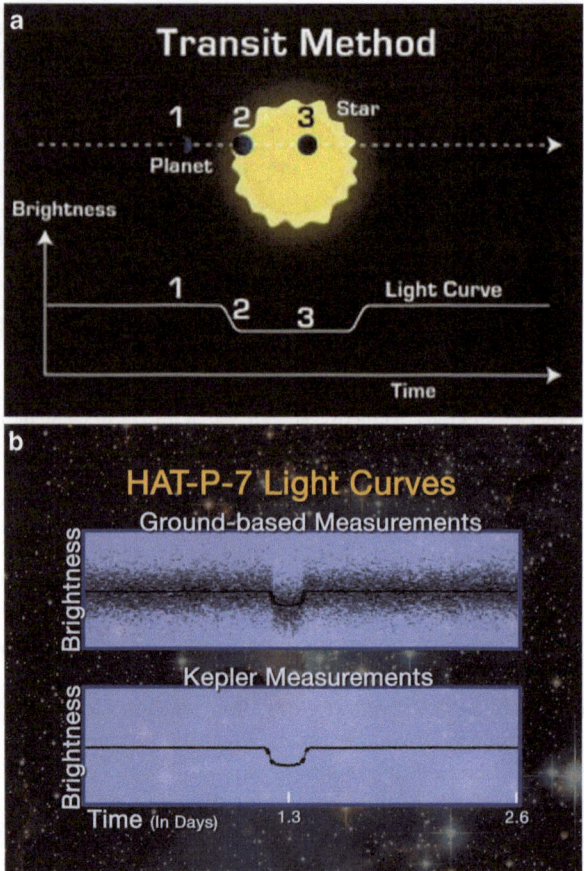

Fig. 1.8 (**a**) Shows how the Transit Method for detecting exoplanets works. Detecting changes in light curves of transiting exoplanets is much easier in space than with Earth-bound telescopes. Our thick atmosphere is filled with constantly moving regions of warmer and cooler air that produces major optical distortion (thermal) of any light beams that travel through it. That is why stars "twinkle" at night, and why any celestial object we are trying to view with a telescope on the ground seems to be constantly quivering or shaking and slightly going into and out of focus. (**b**) Shows photographs of the same exoplanet as it moves in front of (transits) its exoplanet. The *top* photograph was taken with a land-based telescope and clearly shows the poor quality of the image which results from atmospheric thermal distortion, while the *bottom* photograph shows the same event as it was recorded in space by the Kepler space telescope (image credits: NASA/JPL)

Kepler may be a gross under-estimate of the total numbers that might be detectable if Kepler were *able to look* at the **total area** of the sky! Thus, exoplanets may be as common as "*grains of sand in, if not all, at least many of the Earth's beaches*"! Of course, if our own solar system is typical, only one out of eight of these planets may be life friendly, at least for our familiar carbon-based form of life.

Although the Kepler space telescope team (see the NASA 2013 government report entitled "Complete Guide to the Kepler Space Telescope Mission...,"

Fig. 1.9 In June 13, 2013 the planet hunters reported that they had found over 2,000 potential candidates for exoplanets circling other stars, and they *boldly* listed (as well as presented artist's drawings)12 exoplanets that they believed had the highest probability of being habitable. All of these planets are orbiting in their home star's "habitable zone" where surface conditions could possibly be warm and wet enough to support our form of carbon-based life. While larger than Earth and, therefore, classified as "super-Earths", all of these planets would still be potentially habitable (at least for creatures with stronger legs, or bigger wing spans). The astronomers at the University of Puerto (PHL) developed an Earth Similarity Index (ESI) scale based on a number of planetary characteristics including mean radius, bulk density, escape velocity, and surface temperature to rank each exoplanet's similarity to Earth. Any scores between 0.80 and 1.0 are considered to be "Earth-like planets" that could be capable of hosting carbon-based life forms. This stands in marked contrast to the ESI score of 0.16 for Jupiter which indicates this planet is very unfriendly to our form of life. *Asterisks*: planet candidates. Number below the names is the Earth Similarity Index (ESI) (image credits: Planetary Habitability Laboratory @ UPR Arecibo, http://phl.upr.edu, June 26, 2013)

referenced at end of present book) is beginning to identify candidates for possible new exoplanets at an amazing rate, the task of scientifically verifying that they are actually bona fide exoplanets is, and will continue to be for quite some time, a very slow and laborious process. Although the Kepler telescope was launched into space on March 6, 2009, no major exoplanet hunting (or at least reporting) was begun for almost a year. This was because the scientists needed time to test all operating components of the telescope and its land-based support systems to make sure everything was working perfectly. When they did begin serious planet hunting in the spring of 2010, the Kepler team suddenly found themselves being swamped by discoveries of potential candidates for exoplanets. Within just two short years, the scientists had already discovered 1,235 new possible exoplanets, with almost half of these being small enough (Earth size or super-Earth rocky planets) to possibly support some kind of carbon based and water-dependent life. Since scientists are trained to be ultra-conservative and never announce a discovery until they are very confident ("beyond a reasonable doubt") that they are correct, the NASA scientists

Fig. 1.10 Shows a NASA space artist's view of the small area of the Milky Way galaxy that the Kepler space telescope is targeting for possible exoplanets. The targeted area is more towards the center of our Milky Way galaxy and in the mid-plane which contains larger numbers of stars. Portrait of the Milky Way © Jon Lomberg, www.jonlomberg.com (image credit: Kepler/NASA)

did not start making any announcements of confirmed findings of exoplanets until the early fall of 2011. As of March 2014 (the time of the writing of this book), the count for possible new exoplanets has now risen to just under 4,000 (Fig. 1.11a, b) which contrasts with the total of 500 that had been previously discovered by land-based telescopes in the period from 1995 to 2010. And, to stir up the excitement pot a bit more, with the use of the Kepler space telescope which eliminates atmospheric turbulence and allows astronomers to detect smaller planets, our scientists have now started finding exoplanets that are even smaller than our Earth. Until just recently, the smallest exoplanet found orbiting a sun-like star was a rocky world half the size of Earth and almost identical in size to Mars. Although it is probably too hot for life, researchers believe its discovery boosts the chances of finding other, more life-friendly planets. However, on February 22, 2013, the NASA scientists reported that the Kepler space telescope has now found a hot and rocky planet slightly smaller than our own planet Mercury that is circling a sun-like star located about 210 light years away from us.

The amazing succession of discoveries, starting in 1995, of exoplanets orbiting other stars in our galaxy now begs the question of exactly how many potential homes for alien life may be out there. While it is extremely difficult for astronomers to estimate the total numbers of stars just in the Milky Way galaxy alone, the best current estimate is that our galaxy may contain as many as 200 billion stars. With the recent explosion of exoplanet discoveries in the last few years, our astronomers now estimate that there may be at least 50 billion other planetary systems in our

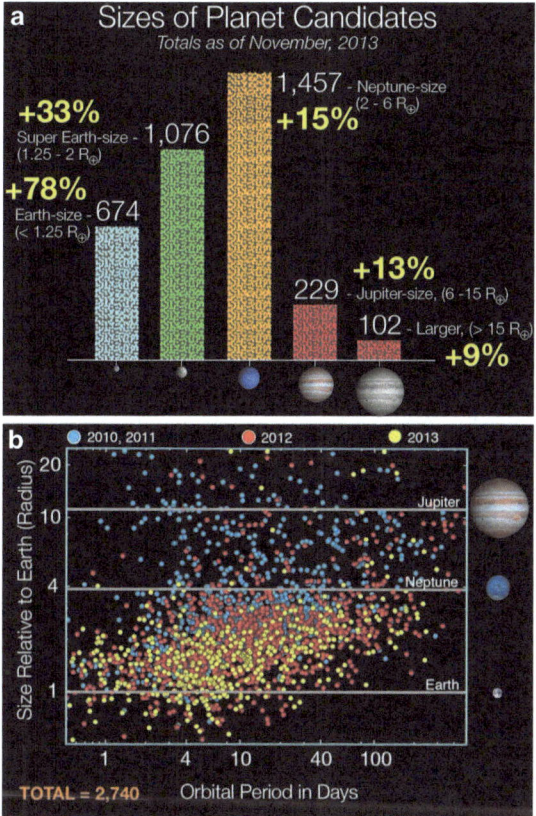

Fig. 1.11 (**a**) Shows the size distribution of the first 4,538 possible exoplanets that Kepler identified in the 4 year period between the beginning of 2010 and the end of 2013 (image credit: NASA/JPL/Kepler). It is apparent that, following the beginning of exoplanet searches in 2010, the highest percentages of new discoveries has started to switch from gas/giant planets to the smaller Earth and super-Earth varieties. Between early 2010 and late 2013, increasing numbers of smaller planets may have started being discovered as a result of improvement in telescope technology as well as observer technique. This suggests that the relative numbers of Earth-size planets in our galaxy may actually be greater than the scientists first believed, and may even be closer to what we see in our own solar system, i.e., four smaller inner rocky planets compared to four larger gas/ice outer planets. (**b**) Shows the size and orbital period distribution of exoplanet discoveries as of January, 2014. The numbers of Earth size and super-Earth sized exoplanets appears to be greater than first thought to be the case although many of the smaller exoplanets do seem to have shorter years (orbital periods) than does the Earth (image credit: NASA Ames Astrobiology Center)

galaxy alone. Since our life scientists believe that liquid water may be a very important, if not critical, prerequisite for the development of life, at least as we know it here on Earth, scientists at both the European Space Agency (ESA) and National Aeronautics and Space Administration (NASA) believe we should give top priority to searching for smaller rocky planets that orbit in their home star's "habitable zones" where the surface temperature and atmospheric pressures would be just right to allow water to remain on the surface in a liquid form most of the time instead of always freezing or evaporating away. If orbiting in a star's habitable zone is a requirement for the development of carbon and water-based life similar to that found on our planet, our scientists now conservatively estimate that, of the 50 billion or more other planetary systems that our galaxy may contain, only 1 % or 300 million might host our kind of life. And, as for attempting to estimate how many potentially habitable planets may exist in our galaxy that might host truly alien life forms that are *not* dependent on carbon and water—Good Luck! One very important lesson that the sudden surge of new exoplanet discoveries by the Kepler space telescope is giving us is that, at least for astronomers, the words "rare" and "common" are virtually worthless with reference to estimating how many "Earth-like" worlds may be out there (Brownlee and Kress 2007; Kasting 2010; Lunine 1999). Such worlds may be extremely uncommon or rare but still number in the trillions in our universe. And, as for estimating the distances our space explorers would have to travel to reach even the closest of these twin Earths—once again, Good Luck! Reality can not only be stranger than fiction, it can also sometimes be very strange!

After 1995, and especially since the Kepler space telescope was launched in 2009, the planet hunters have definitely begun to "open our eyes" with respect to how rare our own home planet may be. Exoplanets that come even close to being twins of Earth appear to be relatively rare in our galaxy, although the fact that we now believe there are so many exoplanets (perhaps 50 billion or more in just our galaxy alone) out there is why this author (and growing numbers of other scientists) are now beginning to occasionally substitute (inappropriately, of course) the word "common" to describe their possible occurrence. Astronomers are now telling us that the universe contains plenty of extreme hostile exoplanets (as defined, of course, by we earthlings). The first series of exoplanets discovered by ground-based telescopes prior to the launching of the Kepler space telescope were, since they produced larger wobbles in the motions of their home stars, massive Jupiter-like gas giants that may have formed much further out in their systems but later migrated into orbits close to their extremely hot home stars.

With improvements in exoplanet detection procedures by land-based and especially the new Kepler space telescope, astronomers are also now beginning to find smaller rocky planets, some of which are close in size to the Earth, that are orbiting even closer to their home stars than our own planet Mercury. One such small exoplanet labeled Corot 7b (which was discovered by the French Space Agency's CoRoT space telescope in conjunction with the European Space Agency) actually completes one orbit of its home star every 20 h! This planet is so close to its star that one side of it constantly faces its star and is extremely hot while the other side is always in total darkness and totally frozen. This planet is believed to have originally formed as a gas giant planet further out in the colder part of its stellar system, but later spiraled in closer to its home star where its outer gaseous layers were burned away leaving a smaller charred rocky core behind. Astronomers have also recently found other Jupiter or Saturn sized gas giants that are orbiting so far away from their home stars that all of their gases (including ammonia and methane) are frozen solid.

Other exoplanets have been discovered that are circling non-sun like stars (e.g. double or triple stars, giants or supergiants, or leftover "monsters" from some of the largest supernova explosions called "pulsars" that are constantly bathing their planets with incredible amounts of deadly radiation). Therefore, our astronomers are now beginning to believe that the universe contains numerous exoplanets that may not be life friendly in any way that the Earth's best life scientists can, at present, possibly imagine (Chela-Flores 2001; Grenfell et al. 2013; Jayawardhana 2011; Schultze-Makuch 2013; Schulze-Makuch and Irwin 2004)! While there may be a relatively small number of exoplanets out there that are friendly to our form of life, there may be many more that are not, *but that might possibly be friendly to other kinds of life that we do not yet know about*! Most of our life scientists now believe that our own home-grown "extremophiles" may provide the first important clues we need to unravel this amazing mystery! So also could the discovery of life elsewhere in our own solar system.

Once Considered Complex and Fragile, Life Now Believed to Also Be Flexible and Resilient

By the end of the twentieth century, the astrobiologists had also started making other even more startling discoveries that now suggests that the form of life (based on the carbon atom and water) that exists on Earth may not be the only form of life that can occur in the universe (Toomey 2013; Ward and Bennett 2008). Our scientists have now discovered that the first small (microscopic) single-cell life forms were able to originate on our planet almost 4 billion years ago when our environment was so hot and hostile that no known forms of living creatures were believed to be possible. The discovery of these **thermophiles** (which scientists call "lovers of hot environments"), whose descendents are still alive today and thriving all over our world, now suggests that life may be far more flexible and resilient than previously thought possible and might easily evolve on other planets that have environments that are totally different from those found on Earth.

In 1977, scientists discovered strange microbial life forms that live deep in our oceans near hot hydrothermal vents. Hydrothermal vents are associated with undersea volcanic activity on the seafloor that allow hot gases and molten materials (lava) to surge up from deep in the Earth into the ocean waters. Scientists from the Woods Hole Oceanographic Institute used a special ocean-going submersible vehicle named "Alvin" (Fig. 1.12), which is about the size of small compact automobile, to explore these vents. Alvin was specially designed to dive to depths of 2 miles or more where the extreme weight of the overlying water would instantly crush a normal submarine. The scientists were diving to a hydrothermal vent system

Fig. 1.12 Shows an artist's drawing of the Alvin deep sea diving submarine used for manned exploration of hydrothermal vent systems, along with a typical collection of black smokers found off the South American coast at a depth of 2 miles below sea level (image credit: David Barczak, University of Delaware)

located close to the famous Galapagos Islands off the coast of South America.[10] The scientists were using search lights attached to Alvin to illuminate any objects they might find near the vents. In addition to a relatively flat ocean bottom littered with scattered rocks and small sea creatures, they found a large number of tall "cigar shaped" black rock-like structures sticking up vertically from the ocean floor. These objects (which are labeled "black smokers") resembled the smoke stacks often seen on factories or homes. These objects were constantly spewing a very hot and dark smoke-like material straight up out of openings located at the top of the structures. A variety of different kinds of sea life were swimming around the black smokers. Some resembled shrimp and crabs while others resembled small fish and mussels. These unusual hydrothermal vents surrounded by large populations of strange sea life have subsequently been found in oceans all over the world. They are located close to the spreading centers of tectonic plates at depths of 1.5 miles or greater below the ocean surfaces. While ocean water at extreme depths is normally relatively cold, with average worldwide temperatures of 3.9 °C (39 °F), the waters around the black smokers of the hydrothermal vents are extremely hot due to geothermal heat coming from the interior of the Earth. It now appears that complex biological ecosystems exist around these vent systems. At the base (i.e., the bottom of the food chain) of these ecosystems are the single-celled heat-loving thermophiles. While the thermophiles themselves may not be dependent on sunlight as an energy source (they thrive on chemical instabilities triggered by geothermal heat

---

[10] The Galapagos Islands are where Charles Darwin discovered how evolution works. Darwin observed how the wildlife fauna, including turtles and small birds called finches, were able to survive on the different islands of the Galapagos chain and very slowly evolve into new species.

when it interacts with specific kinds of high energy minerals such as sulfur), many of the larger (macroscopic) life-forms that depend on their thermophile neighbors as their main sources of daily meals do need the oxygen that is dissolved in the seawater.[11]

The discovery of the thermophiles (including some that even live in boiling water!) in the late 1970s was just the beginning of what has now become a rapidly growing number of other single-cell organisms (plus a few multi-cellular) living on our planet that seem to prefer many other kinds of strange environments that scientists used to believe would be lethal to any known forms of life. Some of these organisms, which scientists now collectively refer to as "lovers of extreme environments" or "**extremophiles**", prefer to live in extremely cold or arid (dry) deserts, in acidic, alkaline, or salty waters, inside frozen ice bergs, in rocks located miles below ground, on top of tall mountains where there is little air and they are constantly exposed to deadly ultraviolet radiation from the sun, and, believe it or not, even on the power rods at nuclear power plants where they are exposed to lethal doses of radiation. Table 1.1 presents a listing of many of the different varieties of extremophiles that I was aware of at the time I was writing this book. Still, by the time the reader sees this book, this list will probably have expanded significantly. It now seems that these organisms can live almost anywhere they choose as long as they have access to water and carbon. And even more bizarre is the additional fact that many of the extremophiles do not need access to sunlight or oxygen in order to survive. *It now seems entirely possible that other exotic forms of life that are based on completely different chemistries from ours may be able to evolve on other worlds in the universe and may actually outnumber our own carbon-based (or even water dependent) form of life*. Thus, with the beginning of the new twenty-first century our scientists, while not yet finding any definitive evidence for the existence of extraterrestrial life (life on other worlds), may be close to discovering that such alien life is far more common out there than we would have dared imagine just a few short years ago.

This recent discovery of extremophiles living everywhere on Earth, while not providing scientific evidence that life exists on other planets, has definitely motivated many of our scientists to believe that extraterrestrial or alien life is far more likely than we previously believed possible. Nevertheless, these strange organisms, in spite of the fact that they are able to live and cope quite well in environmental conditions that the rest of us wimpy earthlings would avoid at all costs, **are not "aliens" in any sense of the word**. They are our relatives. All life on our planet, whether plants or animals, whether bacteria, octopuses, mice, giant red wood trees,

---

[11] While many life scientists now believe that the source of the first carbon-based life on our planet were the hydrothermal vents, other scientists believe life on Earth developed elsewhere. A few scientists, including Jeffrey Bada and Reza Ghadiri, believe that the combination of lightening and hot gases extruded from volcanoes may have fostered the creation of the first life on Earth. The hot materials and gases that belch up from deep in the Earth via volcanic activity are very rich in life friendly nutrients. The interaction between atmospheric lightening and these nutrient materials may have sparked the creation of the first building blocks of life, including amino acids.

Table 1.1 Earth life that has adapted to extreme hostile environments

Extremophile	Description of environment where extremophiles live.
Acidophile	An organism with optimal growth at pH levels of 3 or below.
Alkaliphile	An organism with optimal growth at pH levels of 9 or above.
Anaerobe	An organism that does not require oxygen for growth such as Spinoloricus Cinzia. Two sub-types exist: facultative anaerobe and obligate anaerobe. Facultative anaerobe can tolerate anaerobic and aerobic condition; however, an obligate anaerobe would die in presence of even trace levels of oxygen.
Cryptoendolith	An organism that lives in microscopic spaces within rocks, such as pores between aggregate grains; these may also be called Endolith, a term that also includes organisms populating fissures, aquifers, and faults filled with groundwater in the deep subsurface.
Halophile	An organism requiring at least 0.2 M concentrations of salt (NaCl) for growth.
Hyperthermophile	An organism that can thrive at temperatures between 80 and 122 °C, such as those found in hydrothermal systems.
Hypolith	An organism that lives underneath rocks in cold deserts.
Lithoautotroph	An organism (usually bacteria) whose sole source of carbon is carbon dioxide and exergonic inorganic oxidation (chemolithotrophs) such as *Nitrosomonas europaea*; these organisms are capable of deriving energy from reduced mineral compounds like pyrites, and are active in geochemical cycling and the weathering of parent bedrock to form soil.
Metallotolerant	capable of tolerating high levels of dissolved heavy metals in solution, such as copper, cadmium, arsenic, and zinc; examples include *Ferroplasma* sp., *Cupriavidus metallidurans* and GFAJ-1.
Oligotroph	An organism capable of growth in nutritionally limited environments.
Osmophile	An organism capable of growth in environments with a high sugar concentration.
Piezophile	(Also referred to as barophile). An organism that lives optimally at high pressures such as those deep in the ocean or underground; common in the deep terrestrial subsurface, as well as in oceanic trenches.
Polyextremophile	A **polyextremophile** (faux Ancient Latin/Greek for "affection for many extremes") is an organism that qualifies as an extremophile under more than one category.
Psychrophile/ Cryophile	An organism capable of survival, growth or reproduction at temperatures of −15 °C or lower for extended periods; common in cold soils, permafrost, polar ice, cold ocean water, and in or under alpine snowpack.
Radioresistant	Organisms resistant to high levels of ionizing radiation, most commonly ultraviolet radiation, but also including organisms capable of resisting nuclear radiation.
Thermophile	An organism that can thrive at temperatures between 45 and 122 °C.
Thermoacidophile	Combination of thermophile and acidophile that prefer temperatures of 70–80 °C and pH[a] between 2 and 3.
Xerophile	An organism that can grow in extremely dry, desiccating conditions; this type is exemplified by the soil microbes of the Atacama Desert.

[a]The chemist's concept "pH" reflects the concentration level of hydrogen ions that a liquid solution contains. The range is from 0 (very acidic) to 7 (neutral) to 14 (very alkaline or basic). If the pH value is below 3 or 4 the substance will taste like a raw lemon (very "sour"); if 0 it will have no taste (will taste like water), and if above a level of 10 or 12 it will have what many people call a "soapy" taste

or man are made from the same atoms and share the same basic chemistries (including the same genes and DNA, the same amino acids and proteins, etc.). In spite of looking drastically different on the outside, under the skin (so-to-speak), we all work (function) basically the same.

Since life evolved so quickly on our hostile young world, could it have happened more than once? The discovery of extremophiles living in the absolute worst neighborhoods on our own planet also triggered our scientists to begin wondering about how easy it might be for other kinds of life to pop up on our planet, given that the first life forms may have been able to do so shortly after the Earth first formed. During the first 500 million years following its formation period, the Earth went through an extremely tumultuous time (what geologists refer to as the *Hadean Period* or "heavy bombardment" period) in which the planet was still being frequently bombarded by debris left over from the solar system formation process (Fig. 1.13a, b). Instead of being hit by large objects from space every few million years (e.g. the asteroid strike that 65 million years ago assisted in the destruction of the dinosaurs), during the heavy bombardment period the Earth was still being pummeled by smaller (gravel to boulder size) objects several times daily and by much larger objects big enough to inflict worldwide damage every 100 years or even less. Each time one of these very large objects hit the Earth, enough heat was created to melt the surface of the planet and partially or totally vaporize any early oceans that had managed to form. It would then require some time (possibly a hundred or a thousand years, or even longer) for the water vapor to re-condense from the atmosphere and refill the oceans.

Since we now have evidence that the first fully formed single-cell organisms were present on Earth possibly as early as 3.8 billion years ago, or even earlier (Dreamer 2011; Schopf 2002), some scientists now believe it is possible that other forms of life might have managed to get started on our planet even earlier, perhaps during the tail end of the late heavy bombardment period, but were wiped out by the still continuing bombardment activity. Life might have experienced many repeated "false starts" followed by extinctions during this traumatic time in Earth's early history. Some scientists now think it may be possible that some organisms in some of the later false starts might have managed to survive, perhaps because they were located in deeper portions of the oceans or perhaps deep below ground. If so, some of these creatures might still be alive and well hidden somewhere on our planet where they would have been lucky enough not to be eaten by our modern animal species. Thus, it is possible that the genesis of life on our own planet may have occurred more than once.

Most life scientists believe any systematic search for a second independent origin (genesis) of life on Earth, or what life scientists label as a "shadow biosphere" (Toomey 2013) will be extremely difficult for two reasons. First, these creatures would probably be "hiding" in locations on Earth that would be very difficult for our scientists to physically access. They would have needed to arise in locations (e.g., deep below the Earth) where they could survive the catastrophic bombardments. And, secondly, they would probably be difficult to identify using

Fig. 1.13 (**a**) Artist's view from space of the extremely hot Earth about 4 billion years ago when it was being pummeled by leftover debris from the solar system formation period. In spite of the incredibly hot surface conditions, many scientists now believe the first single-celled microbes that could withstand heat were able to evolve (image credit: NASA). (**b**) Space artist's drawing of what the surface of the Earth may have looked like during the intense early bombardment period following its formation. Our world would have been the target of constant impacts from large leftover debris (meteors, asteroids) which would have rendered our early oceans hotter than the hottest of today's sauna baths (image credit: David A. Hardy/www.astroart.org)

our scientists (biochemists) traditional chemical analysis procedures, since almost all of our procedures have been specifically developed to find carbon-based life "as we know it". However, many life scientists now believe this search for a second genesis of life, or at least a drastic alternative type of biochemistry on Earth, is definitely worth doing. In the last few years, more and more of our diehard astrobiologists have begun climbing down into deep caves (or conning diamond mine companies into allowing them to poke around in their deepest mine shafts), tunneling deep inside large frozen icebergs, or hitching rides on deep sea submersibles to search for weird organisms in our deepest oceans.

In late 2010, a research project sponsored by NASA was reported that initially appeared to have possibly "hit paydirt" in this search for alternative life chemistries on our planet (Fig. 1.14). Earlier that year scientists began exploring Mono Lake in

Fig. 1.14 In 2010, scientists unexpectedly discovered a polluted lake in California that has been isolated from sources of fresh water for more than 50 years that has high concentrations of the chemical arsenic. This lake appears to host an unusual type of single cell microbe (see *inset*) that may possibly substitute the poison arsenic for phosphorus in constructing its genetic materials (RNA, DNA). Whether this unexpected finding represents a mutation of our familiar carbon-based form of life or is the first discovery of a second independent genesis of life on our planet is now being investigated (image credit: NASA)

California which has an unusual chemistry as a result of its being isolated from any sources of fresh water for over 50 years. The lake waters have very high levels of salt, alkaline, and arsenic. On Earth, the six atomic elements that constitute the basic building blocks of all known forms of life (from microbes to red wood trees to man, etc.) are hydrogen, carbon, nitrogen, oxygen, phosphorus, and sulfur. The microbes that were found in Mono Lake are a strain of microbe called GFAG-1 that is similar to a common group of bacteria called "Gammaproteobacteria". GFAG-1 is now the first organism that has been found to break the rules in being chemically different from all other forms of known life on our planet. Initial studies indicated that GFAG-1 has substituted the element arsenic for phosphorus as a part of its vital biochemical machinery, including DNA and RNA, the structures that carry the genetic blueprint for building life which is critical to all living cells. More recently, however, scientists have started to question whether this organism does incorporate arsenic into the actual chemical structure of its genetic materials or just simply relies on it for its normal growth. Thus, it is now unclear whether this organism constitutes a descendent of an earlier second independent chemical genesis of life on Earth or is the result of some kind of unusual evolutionary mutation of our familiar form of life. Still, if nothing else, the discovery of this strange organism raises the bar for the amazing diversity of life on our planet to an even higher level (notch). It definitely provides fuel for the growing belief of our astrobiologists that life can arise quite easily on Earth as well as elsewhere in the universe, and could be chemically quite diverse.

Could life have developed on other moons or planets in our solar system? The discovery that some of our fellow non-human relatives prefer to live in the most hostile and unfriendly places on our planet, plus the very recent discovery of strange microbes living in arsenic filled lakes in California, reinforce our scientists' belief that life may be tough enough to be able to easily pop up in all kinds of strange environments throughout the universe. In order to investigate whether life actually is common in the universe, our scientists may not need to wait until mankind is able to make personal trips (or launch robotic missions with advanced remote sensing technologies) to other interstellar locations (other exoplanets). It is entirely possible that, in the next century or two, our astronauts will travel to other locations in our own solar system and discover the existence of totally different life forms (other independent origins or geneses of life). ET may not be all that far away from us (Chyba and Phillips 2007; Cohen and Stewart 2002; Jones 2004; Jakosky 1998; Jakosky et al. 2007; Lunine and Butler 2007)!

Until close to the end of the twentieth century, most scientists believed that the possibility of viable life-forms anywhere beyond the orbit of Mars in our solar system was small to nil. The reason for this belief was simply that, in addition to being far too cold (even water is frozen solid), there were very few known environmental conditions (e.g., solid surfaces for animals to walk or crawl on) anywhere that could possibly allow carbon/water based life, as we know it, to exist. Nevertheless, the story may be different for at least two of the inner planets. Two of the first places we will definitely search for ET will be Mars (Jakosky et al. 2007) and Venus. It is interesting that some scientists believe we have already found evidence for microbial forms of life in meteorites that are thought to have been delivered from Mars to Earth millions of years ago and may still be alive today on Mars and busily exhaling methane gas as a metabolic waste product (Schulze-Makuch and Darling 2010). While Venus (Fig. 1.15a, b) is definitely too hot today (temperature close to 900 °F) to host life on its surface (or at least life "as we know it"), our sun was much cooler about 3 or 4 billion years ago and the surface of Venus may have been cool enough for at least a billion years after it formed to host life friendly oceans (Fig. 1.16). Some scientists believe that Venus, at this early time, might have been even more likely than Mars to have first jumpstarted life in our solar system. In spite of Venus today being in the throes of a geologic runaway greenhouse event, a few researchers believe it is possible that some microbes from that earlier more temperate period might have survived and evolved to today be doing quite well in Venus' cooler upper atmosphere. In contrast, most astrobiologists today believe Mars was, some 3 or 4 billion years ago, covered by large bodies of water and hosted an atmosphere and environment that was possibly quite friendly for at least the beginnings of single-cell microbial life (Fig. 1.17a, b). The descendants of many of these creatures may today still be living quite comfortably in warm and wet underground homes on Mars and waiting for our arrival (Jakosky et al. 2007; Sullivan and Baross 2007). In 2003, NASA scientists used infrared

Fig. 1.15 The planet Venus is today far too hot to sustain any kind of life that we know about. (**a**) Shows a recent NASA photograph that shows what the planet looks like today covered by a very thick cloud cover, while (**b**) shows an artist's drawing depicting that the surface of Venus may today be extremely hot and covered by intense volcanic activity (image credit: European Space Agency)

spectrometry to obtain evidence that, at certain times of the year, larger than expected quantities of methane gas appear to be released into the Martian atmosphere. This seasonal effect suggests the possibility that primitive life forms, similar to some species of bacteria on Earth that toss out (eliminate) methane as a metabolic waste product instead of CO_2 or O_2 (as do our animals and plants, respectively) may reside in warmer and wetter subsurface locations on Mars today.

While the giant gas planet Jupiter, which is even further out in the colder distant part of our solar system, has a total of 63 moons (at last count), the only ones believed to possibly be life friendly are three of the four largest moons (i.e., the Jovian or Galilean moons) that are large enough to have been discovered by Galileo Galilei with his first telescope in 1610 (Fig. 1.18). These moons have orbits around Jupiter that are slightly less than circular (i.e. more oval shaped than round) which means that their distance from Jupiter is constantly changing as they orbit their

Fig. 1.16 Many scientists believe that Venus may not have always been the hot and life hostile planet that it is now. This image prepared by Reuben Reyes and his students indicates that things might have been totally different early in this planet's history. Most astronomers believe that, since the normal evolution of stars involves a very slow warming trend with age, 3 or 4 billion years ago our sun was not as hot as it is today, and Venus could have been much cooler and may have actually harbored lakes or oceans filled with liquid water that could have supported the early stages of the development of at least microbial life. Dr. Reyes' developed a very interesting computer-generated image which depicts what early Venus might have looked like covered with vast oceans about 1 or 2 billion years following its Initial formation. The Reyes team used the radar imaging data collected by the 1989 NASA/JPL Magellan space probe to obtain a detailed "elevation map" of the entire surface of Venus and then digitally inserted a computer generated ocean that could have covered one-half (50 %) of the total planetary surface (image credit: Reuben Reyes, University of Texas/Austin Department of Astronomy)

home planet. This causes them to be engaged in a small gravitational tug-of-war effect with Jupiter which results in the interior of each moon being constantly stretched back and forth to create internal frictional heat (which astronomers refer to as *Tidal heating*). The closest moon, Io, has the greatest amount of tidal heating which causes its surface to be hot and molten with frequent volcanic activity. Io is probably too hot to support life The second moon out from Jupiter, i.e., Europa, has considerably less tidal heating than Io but is now believed to possibly have enough to host relatively warm and life friendly salt water oceans below its thick ice covered surface (Fig. 1.19a). The two more distant Jovian moons from Jupiter, i.e., Callisto and Ganymede, might also have underground salt water oceans that could be warm enough to host some forms of life (Fig. 1.19b). The ocean floors of these three Jovian moons, because of this tidal heating effect, may host hydrothermal vent systems that could give rise to heat-loving thermophiles not unlike those found in the vicinity of the hydrothermal vents or black smokers located on the

Fig. 1.17 (**a**) Shows an artist's drawing of what Mars may have looked like, circa 3 or 4 billion years ago (i.e., at the same time as Earth's late Hadean or early Archean Eras). The planet at that time is believed to have hosted a thicker atmosphere and lakes or small oceans on its surface and a hotter internal core that supported a thicker atmosphere that could have allowed the evolution of microbial life. (**b**) Shows a photograph of today's much drier and probably colder Mars that has lost much of its atmosphere (image credits: NASA)

floors of Earth's deepest oceans. Some scientists, however, including Christopher Chyba, believe that the extreme distance of the Jupiter system from the sun, which results in extremely low freezing temperatures (-260 °F or below) on the surfaces of the three outer Jovian moons might preclude there being enough heat from any hydrothermal vents to support life. However, Jupiter is unique in having a large and very strong magnetic field that produces a constant pounding of Europa's outer ice

Fig. 1.18 Shows NASA photographs of Jupiter and its four largest moons as they would appear if lined up for a family portrait. The four moons, from *top* to *bottom*, are Io, Europa, Ganymede, and Callisto (image credit: NASA)

layer by fast-moving electrified atomic particles (protons, electrons) from space. Europa lies deep within this strong magnetic field and is continually bombarded by billions of these atomic particles every second.[12] These particles, or ions as they are referred to by chemists, could cause chemical reactions to occur in Europa's ice layer that might transform water and carbon dioxide into new organic compounds such as formaldehyde (CH_2O) and other carbon-based organic fuels that could become dissolved into the ocean waters to support the basic chemical reactions needed to foster life (Chyba and Phillips 2007). Single-cell microorganisms (bacteria) that rely on formaldehyde and other complex organic molecules as their primary food sources are common in the Earth's oceans and might also be present on any moons in the outer solar system that are warm enough (or become warm enough in the far distant future) to support liquid water.

And, even further out in our solar system, the ringed planet Saturn has a relatively large moon named Titan that is actually covered by a gaseous atmosphere that is even thicker than Earth's atmosphere which contains high concentrations of organic molecules (i.e., hydrogen and carbon bonded to each other and other life friendly elements including nitrogen). This moon is so cold, with an average surface temperature of −290 °F, that all of its surface water is probably frozen rock solid

---

[12] The amount of radiation hitting Europa every day is 540 rem which would be fatal to unprotected humans and is ten times greater than that found in the Earth's van Allen radiation belt. Thus, what is "bad" for humans could be "good" for bugs living on Europa. Score another point for the possible extreme diversity of life in our universe!

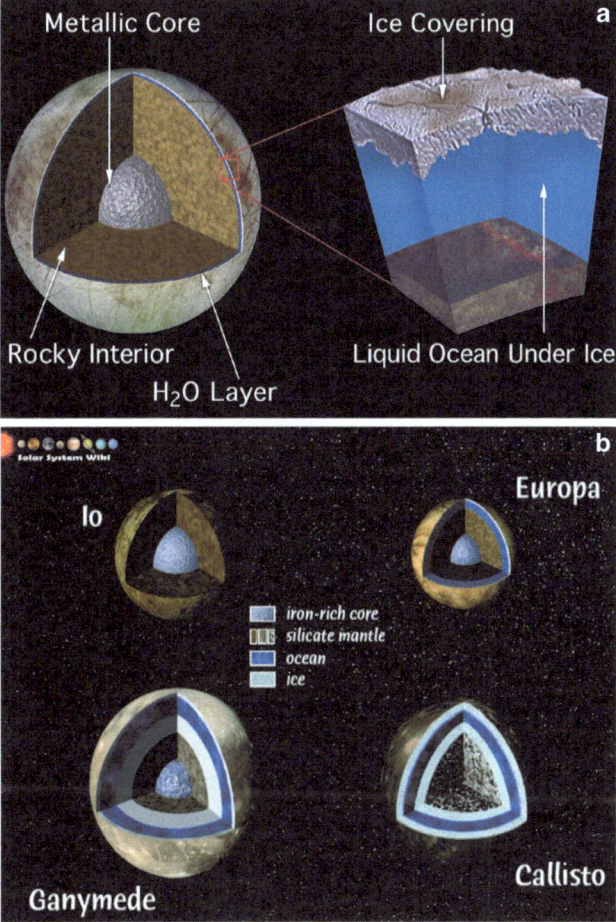

Fig. 1.19 (**a**) Europa is believed to have a large saltwater ocean below its thick ice covered surface that may receive enough heat from the interior of the moon to possibly host some forms of life. Although Ganymede and Callisto may not have ice covered surfaces, some scientists believe they both may (**b**) host small oceans below their rocky surfaces that could be life friendly. All four of the larger inner moons of Jupiter orbit Jupiter in slightly non-circular paths that produces a constant gravitational tugging effect with Jupiter that causes their interiors to be slightly heated (like a paper clip being constantly bent back and forth until it breaks) (image credits: NASA)

and would not be available to serve as a solvent that could support any Earth-like forms of biological chemistries. Of course, Titan does support rivers on its surface that flow into lakes or small oceans containing liquid methane and ethane (Fig. 1.20a, b) that might support some other kinds of carbon-based life (Lunine and Butler 2007). The fact that Titan appears to have active volcanoes that have provided a source for this moon's rich gaseous methane atmosphere also suggests that life forms on Titan might be really strange creatures that actually survive by inhaling (breathing) hydrogen rather than oxygen. Nevertheless, in March of 2012

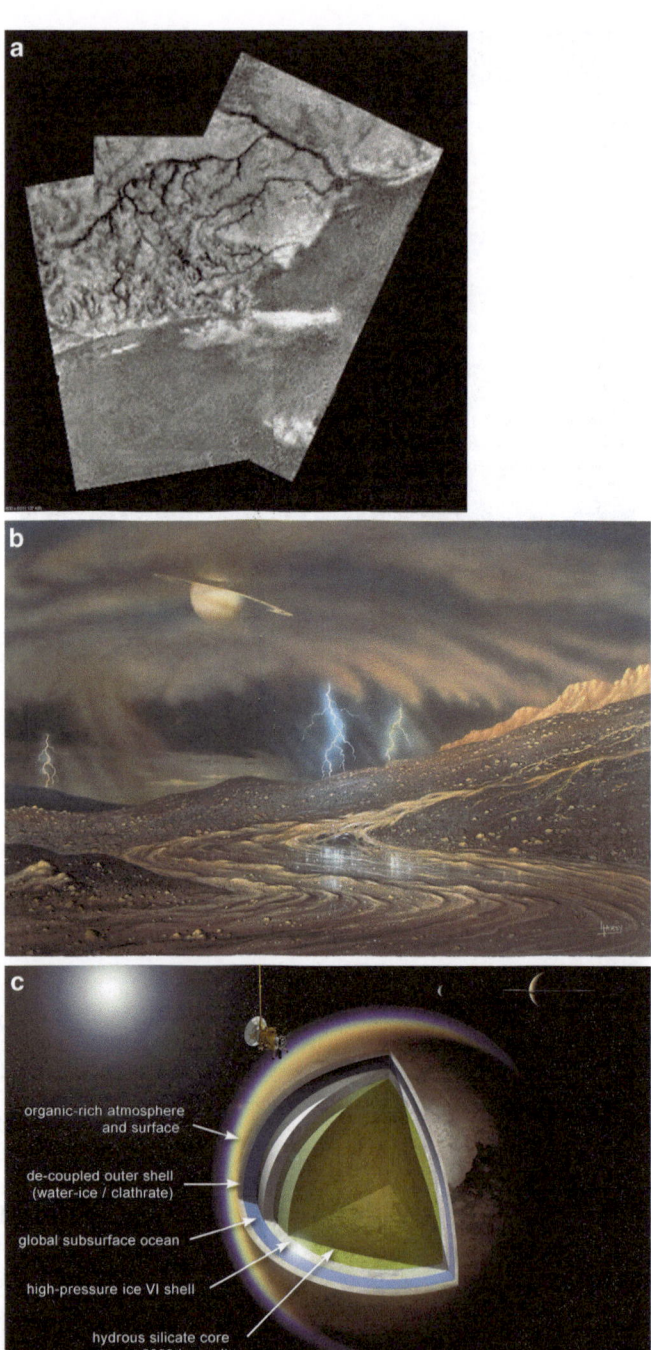

Fig. 1.20 NASA scientists now believe that Saturn's moon Titan may actually host liquid oceans on *both its surface* (filled with liquid methane and ethane) *and in underground locations* (possibly filled with liquid water). (**a**) Shows a photograph taken by the Cassini Huygens probe as it

NASA and the Jet Propulsion Laboratory (JPL) issued a joint news release that the NASA Cassini spacecraft has found evidence that Titan may also host a slightly warmer (due to tidal heating) underground ocean of liquid water (Fig. 1.20c), which opens the door to the possibility of the presence of ocean species that might not be entirely unlike our own carbon-based ocean life.

Thus, in spite of its extreme cold environment, Titan is one of the most complex and interesting chemical environments in our solar system, since it is constantly producing large varieties of different organic molecules that might allow our life scientists to solve the mystery of how life may have originated on our own planet. The presence of methane in both gas and liquid forms on Titan may provide clues as to how these life critical chemicals might be converted by sunlight (and cosmic radiation) into more complex organic precursors to life. Scientists have discovered the existence of a unique family of organic molecules on Titan named *tholins* that are believed to possibly be the precursors to carbon-based life as we know it on Earth. Tholins are large organic molecules that are believed to be formed when ultraviolet light from the sun (or any nearby stars) interacts with mixtures of more simple gaseous forms of organic molecules containing hydrogen, carbon, and nitrogen to produce new and different kinds of molecules containing these same elements (Fig. 1.21). Tholins are generally reddish-brown (or "muddy colored" according to Carl Sagan). The haze and orange-red color of Titan's atmosphere are thought to be caused by the presence of tholins. Complex organic molecules similar to tholins have also been found in many other places in the universe including some of the other icy moons in our solar system as well as the icy surfaces and gaseous tails of comets plus the protoplanetary disks of young stars in distant gas and dust clouds such as the Orion Nebula where new stars and planetary systems are being born. Several scientists have speculated that our own planet may have been seeded by organic compounds early in its development by tholin-rich comets. Tholins do not presently exist on Earth because the large amounts of oxygen that is present in our atmosphere prevents their chemical formation. These chemicals can only be formed in an atmosphere, such as Titan's current oxygen-free atmosphere or Earth's atmosphere that existed about 4 billion years ago before certain kinds of bacteria (cyanobacteria) and plants started producing oxygen as a metabolic waste product and releasing it into the atmosphere. However, tholins can be created and studied by scientists in Earth-bound laboratories. It is believed that these life friendly compounds, which are organic aerosols, i.e., particles small enough to remain suspended in the atmosphere for some time, are derived from solar irradiation of

← ───

Fig. 1.20 (continued) descended by parachute to Titan's surface that show rivers flowing into large lakes, while (**b**) Shows an artist's drawing of what the surface of Titan might look like close up. While all surface water on Titan is frozen solid, these small rivers and lakes on the moon's surface appear to be "warm" enough to be filled with liquid methane and ethane. This, of course, means that when precipitation falls from the Titan clouds, what you get is huge fluffy "snow flakes" or large "rain drops" (which can be golf ball size or even larger) made of methane rather than water. However, (**c**) shows that if you dig deep enough below Titan's land surface, you may find oceans (depicted in *blue color*) that are warm enough (due to Tidal heating) to allow liquid forms of water to exist (image credits: NASA/JPL)

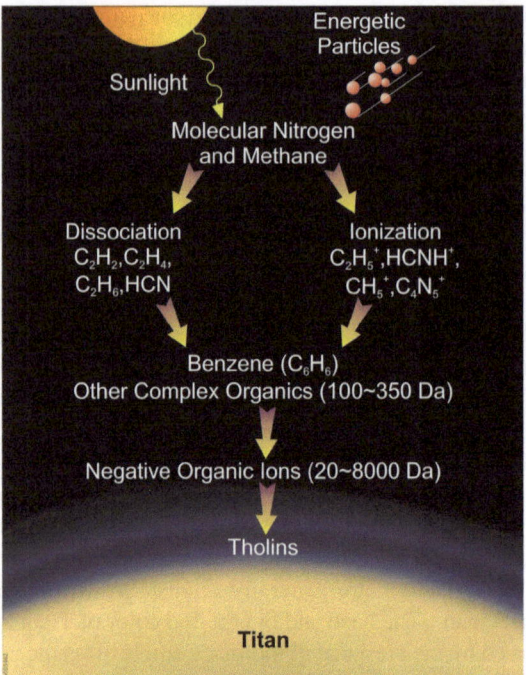

Tholin formation in Titan's upper atmosphere

Fig. 1.21 The atmosphere of Titan contains large quantities of molecular nitrogen (N_2) and methane (CH_4) gases. When sunlight from nearby stars (e.g., our sun) and/or high energy atomic particles (e.g., protons, electrons, neutrons, neutrinos) collide with these nitrogen and methane gas molecules they cause them to be broken apart, i.e., *dissociated*, into two or more smaller parts, or undergo *ionization* which is a chemical process by which electrically neutral atoms or molecules are converted to electrically charged (positive or negative) atoms or molecules. These dissociated and ionized atomic particles are then linked together with each other to form larger and more complex organic compounds including *tholins*. Some scientists believe these complex organic compounds (tholins) are very common in other parts of our own solar system and may have played key roles in the development of carbon-based life not only on Earth but elsewhere in our solar system as well as other parts of the universe. This suggests that carbon-based life may be relatively more common in the universe than many scientists now believe is possible (image credit: Wikipedia Commons)

mixtures of gaseous nitrogen (N_2) and methane (CH_4) in the atmosphere of present-day Titan. Thus, tholins contain many of the chemical precursors to life that could eventually lead to the production of real bona fide amino acids, which are the building blocks of life's proteins on Earth plus perhaps numerous other locations throughout the universe. Tholins may have interacted with liquid water in a process called hydrolysis to produce the chemical components of early life on Earth. If liquid water is not already present in warmer underground locations on present-day Titan, it might become available in a few billion years when our sun warms up Titan enough to allow the existence of liquid water on its surface.

Fig. 1.22 While Jupiter's moon Europa (**a**) today has a very large liquid water ocean that is completely covered by a very thick ice cap, this could possibly change in another 3 or 4 billion years as the sun continues to warm up and enters its red giant stage (image credit: NASA). (**b**) Shows one space artist's view of what the surface of Europa might possibly look like at that time. Europa's entire surface could be covered by a deep water filled ocean that might host advanced forms of sea life (image credit: Calvin J Hamilton)

The recent discovery of tholins, therefore, raises the important question of whether the increased warming of Titan by our sun that is expected to occur in the next few billion years might lead to the rise of liquid water on this moon's surface that could support the evolution of carbon-based life forms not unlike our own? And what might similar solar warming do with respect to enhancing the evolution of carbon-based life on any of the other icy moons in our outer solar system (or similar moons of any giant gas planets of other solar systems that are endowed by the presence of life friendly atomic elements, e.g., hydrogen, carbon, nitrogen, etc.)? In another 3 or 4 billion years, mankind (unless we manage to escape) will be no more than piles of burned out carbon residues on a very hot charred planet, while newly formed water-filled oceans on Titan, and also possibly Europa (Fig. 1.22a, b), and some of the other former icy moons of our outer solar system may begin to be filled to the brim with playful fish-like animals or other forms of sea life that may be vaguely similar to those that some of the inner planets of our solar system hosted many eons earlier. Thus, in a very real sense it may turn out that the apparent universal tendency of stars to slowly warm up over time as a part of their normal evolutionary trek may make for a very interesting "traveling road show" with respect to the evolution of ETs on other worlds all over the universe that like it wet and warm in order to prosper. Two or three billion years ago, Venus might have been cool enough to host warm oceans and carbon-based sea life. In more recent years it has become Earth's turn to support such life forms, and in another 1 or 2 (or 3) billion years our oceans may boil away and our form of life may disappear. In the meantime, the icy moons of the outer solar system may warm up and take over the job of hosting large varieties of carbon-based sea life.

Photographs in the early 1980s by NASA's Voyager mission to Saturn revealed the existence of a second smaller and quite cold moon of Saturn named Enceladus (surface temperature of $-200\ ^\circ$C, or $-330\ ^\circ$F) that appears to have a young smooth surface in which old impact craters are continually wiped out by water that is regularly deposited onto the surface to form an ice covered moon. In 2010, the NASA/ESA Cassini mission to Saturn confirmed the existence of at least 90 jets or fissures of all sizes that are spewing huge geysers high into space from pockets of water located below the South pole region (Fig. 1.23a, b) of this moon. NASA has now confirmed that these geysers contain large quantities of salty water vapor, carbon dioxide, and organic molecules (including tholins) which suggests that this moon, like Jupiter's Europa, may host a large saltwater ocean below its ice covered surface that could support carbon-based life forms similar to those that exist in our own oceans as well as possibly those of Europa (plus Ganymede and Callisto). It is believed that Enceladus, similar to the Jovian moons of Jupiter as well as Saturn's moon Titan, has a slightly non-circular orbit around its home planet (Saturn) that causes a tidal heating effect. The oceans of Enceladus, like those of Jupiter's Europa, may be warm in spite of this moon's surface being frozen solid and incredibly cold. Could these oceans have already spawned some primitive microbial carbon-based life forms, and what will happen as the oceans warm up further in the next few billion years?

Although Europa will, in the next few billion years as the sun becomes a red giant star and warms up, experience a total meltdown of its ice caps and become a water world, what will happen to Saturn's much smaller and more distant moon Enceladus is debatable. Both Europa and Enceladus currently do have atmospheres. Europa has a thin atmosphere composed of non-biological produced oxygen that results from charged particles from space breaking its water vapor into hydrogen and oxygen atoms. Enceladus, in contrast, appears to support a "significant" atmosphere composed mostly of water vapor plus small quantities of carbon dioxide and methane gases. Both moons are very small with Europa being slightly smaller than Earth's moon, while Enceladus is only about 1/7th the size of our moon. If the ice caps of these two moons do melt in another few billion years due to our sun warming up, the fact that Europa is much closer to the sun than is Enceladus as well as being considerably larger then Enceladus may allow this moon to sustain surface water oceans for a much longer period of time that could possibly support the evolution of more advanced forms of sea life (see Fig. 1.22b). While Enceladus' greater distance from the red giant sun might protect it somewhat from heating up,

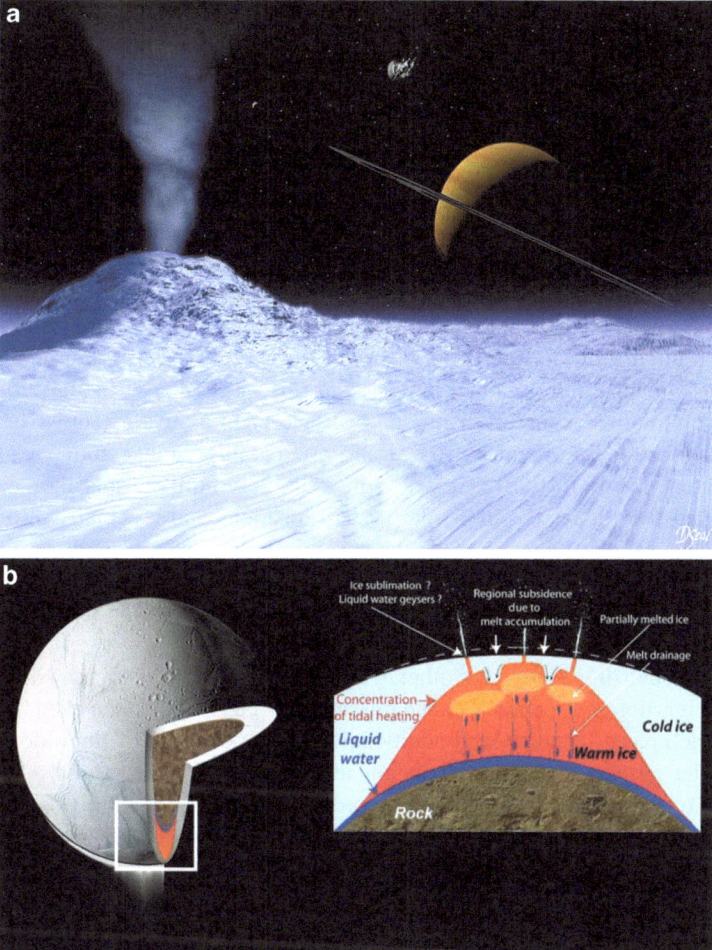

Fig. 1.23 (**a**) The moon Enceladus in the Saturn system has been found to spew hot steamy geysers from its interior that contains water and certain organic molecules that could be associated with some form of subsurface microbial lifeforms. (**b**) Shows an artist's drawing of how these water geysers may be created (image credits: NASA)

its smaller size and lower gravity would likely result in any oceans (or lakes) evaporating more quickly.

Therefore, in the new twenty-first century, our scientists are suddenly beginning to find new evidence that other forms of either single-cell or even primitive multicellular life may exist even in our own backyard, i.e. our solar system. **And, as unbelievable as it may now seem, it is beginning to look like the moons of planets and not the planets themselves, may be the best places to look for such alien life forms** (Coustenis and Blanc 2012; Scharf 2008; Chela-Flores 2013;

Lunine and Butler 2007).[13] No scientists, however, today believe that multi-cellular life forms anywhere close to man's level of complexity exists in our solar system. In contrast to any life forms we may discover living on far distant exoplanets located many light years away, some of these alien "neighbors" in our own backyard may not, however, be independent geneses (origins) of life but could actually be our "distant relatives".

For many years, a few scientists have believed that life might be able to arise in one location in a planetary system and later be transported alive to other planets by meteors or small asteroid collisions. Although this idea of *panspermia* has received no substantial scientific support, it does have a small cadre of supporters who suggest it could be possible for primitive life (e.g., single-cell microbial life forms) to hitch a ride inside rocks or other debris blown away from one planet's surface (by an asteroid or comet impact) and survive the long trip and subsequent landing on another planet's (or moon's) surface. While most scientists believe it might be possible for organisms to be transported in this manner from one planet to another (e.g., Mars to Earth), it would be extremely unlikely that they could be transported from inner to outer solar system locations and especially difficult for such "refugees" to escape or make their way into any underground (or under ice locations) such as the moons of Jupiter and Saturn. Still, if this theory does turn out to be true, and we do find life on Mars, that life may be our long lost cousins. We may all be Martians (or Venusians?).

Even though the search for other ETs in our own solar system will be quite difficult, it is not impossible. Manned missions to Mars may happen even before the end of the present century, and additional missions to some of the moons of the outer solar system may possibly begin as soon as the next century. However, manned missions (or even unmanned probes) to other potentially life friendly worlds elsewhere in our vast galaxy or even beyond will likely remain only a dream for our frustrated scientists for quite some time. However, the good news is that, as we will now describe in the next chapter, science is now beginning to find definite evidence that the "chemical stuff of life" appears to be out there, if only we can find a way to reach and study it.

---

[13] A few scientists have even suggested that some of the largest of the objects in our asteroid belt (e.g., Ceres which has a diameter of 590 miles) may host small subsurface salt water oceans that might contain tholins and be warm enough to allow the development of carbon-based microbial life forms.

Chapter 2
When Will Mankind Achieve *"First Contact"* with Extraterrestrial Life Forms

It now appears very unlikely that our astronauts will encounter any kinds of life in our solar system that are of the intelligent persuasion anywhere close to our Earth-bound multicellular animal or even human types. However, as I indicated in Chap. 1, many of our space scientists now think it is possible, and a few believe even likely, that early stage primitive forms of life may have been able to evolve in the earlier histories of some of our inner planets (e.g., Venus and Mars) or on some of the moons of the outer planets, and some of these creatures may still be alive today. Although single-cell or microbial forms of life might still be present today in some locations, no multicellular forms are likely, assuming, of course, that life elsewhere is based on something equivalent to our "cells". If man wants to search for life forms that are anywhere close to being like us in terms of being intelligent and even capable of communicating with us in some manner, we will need to extend our search for ET to other worlds outside our own solar system. At least for the near future, rocket science will not help us very much with this task. Mankind will need to become incredibly patient and somehow learn to cope with the real possibility that it may be our great-great-great…grandchildren and not us that will first view the exciting pictures and other data/information that our incredibly fast unmanned inter-stellar rocket ships will beam back to us sometime in the far distant future. This would, of course, assume that the speed of light is our ultimate speed barrier, or that nothing like "worm holes" or other fancy future technologies could be developed that might make it possible to somehow cancel or alter what appears to be a insurmountable and universal time/space restraint.

There appears to be another "universal" truism or law of nature that would also preclude us from searching for intelligent forms of life in our own solar system and that is that the evolution of intelligent life, in addition to probably being a very rare event in the universe, is also incredibly slow. If it turns out that humans are the only

© Springer International Publishing Switzerland 2015
J.L. Cranford, *Astrobiological Neurosystems*, Astronomers' Universe,
DOI 10.1007/978-3-319-10419-5_2

advanced intelligent life forms that exist in our solar system, we must look beyond our own "backyard" to even have a chance at finding other life forms that have had enough time, as well as luck (in terms of avoiding destruction) to be able to slowly evolve into creatures more similar to us. Fortunately, the recent amazing discoveries that homes for extraterrestrial life may be common elsewhere in the universe and that life may be both diverse and quite resilient, is triggering many of us (scientists as well as non-scientists) to pose the profound question of how soon it will be before we make the giant leap from speculating about extraterrestrials to being able to actually observe them close up and perhaps develop some form of social relationships. Therefore, it now appears that, if we want to hunt for more intelligent forms of life, we must focus our searches on those incredibly vast and hostile regions of space that lie beyond our own little corner of the universe. Thus far, even the closest exoplanets that are believed to be possible candidates for harboring life are located so far away (the closest one found to date is 20 light years away) that even our best and most optimistic rocket scientists doubt we will be able to visit them anytime soon.[1] And our astronomers tell us that the largest and most powerful land-based telescopes we currently have, as well as the even larger systems including space telescopes that are planned for the near future, will probably be inadequate (for many years to come) to allow our astronomers to visually examine an exoplanet's surface for any physical signs (e.g., cities, large engineering projects) of the presence of advanced technological civilizations. What makes the direct viewing of exoplanets so difficult is that they are all located extremely close to their incredibly bright home stars. Many stars are at least 1,000,000,000 (one billion) times more bright than even their larger exoplanets, and 1,000,000 times brighter if just the infrared portion of the light spectrum is measured. The task of viewing exoplanets, therefore, is comparable to trying to use a telescope to look at a small insect (housefly or bee) flying a few feet away from a powerful search light that is located many miles away. Nevertheless, our computer scientists, astronomers, and rocket scientists now believe this task is not impossible but may actually be accomplished, if not by the end of the present century, then perhaps sometime in the next century or two. The sad reality is that our scientists may presently be very close to possessing the necessary technology to accomplish this exciting feat if only the monies to pay for it could be found. It would likely cost no more than a small fraction of the world's combined military (defense) budgets for a single year. Having expressed this somewhat depressing thought, I will now

---

[1] In early October, 2012, a team of astronomers led by Stephane Udre and Xavier Dumusque used the 3.5 m telescope at the European Southern Observatory to verify that the closest star to our sun, Proxima Centauri, which is located only 4.35 light years from the Earth, hosts an exoplanet. This planet is similar in size to Earth, orbits its home star every 3.3 days and, therefore, has a mean surface temperature of 2,240 °F which would definitely not be friendly to carbon-based life forms like us. However, astronomers are now intently studying the Alpha Centauri triple star system for possible evidence of other exoplanets that might be orbiting in cooler life friendly habitable zones.

proceed to describing how our diehard scientists plan to accomplish this exciting and important task if the cost issue can be resolved.

What Kind of Telescope Could View Small Dim Exoplanets Located Close to Extremely Bright Stars

If we want to conduct searches for real living creatures on other worlds, we have to do much more than just look for dips in the brightness levels of distant stars that transiting exoplanets produce, or minute changes in color (light Doppler shifts) associated with stars being made to wobble by their larger exoplanets. We must look through the clouds and hazy atmospheres of the exoplanets themselves and look for visual or spectrographic evidence of the presence of living creatures that are busy polluting or otherwise altering their planets' surfaces in such a way that can only be explained by the presence of industrious life forms. Since our turbulent and increasingly polluted atmosphere is the astronomer's single greatest impediment in using telescopes to observe distant exoplanets, one solution to this problem would be to place telescopes containing extremely large optical systems (e.g., mirrors for collecting light from distant stellar systems) into space or to build them on the surface of the moon. However, the problem is that telescopes big enough to allow us to see distant planets in sufficient detail to determine whether life might be present, would be truly ENORMOUS! Telescopes of the necessary size to accomplish this feat would be, for the foreseeable future, far too expensive to construct.

Another less expensive solution would be to launch telescopes into space that contain special *star shades* that could block out most of the home star's light and allow the telescope to focus specifically on the light being emitted by the exoplanet itself. NASA has already designed such sun-block systems that could, if funded, be built and placed into space to view exoplanets. Figure 2.1 shows an artist's drawing of what one version of this type of telescope might look like. This telescope utilizes an internal light blocking technique or device known as a *coronograph* that was originally invented by astronomers to block out the glare of the sun's surface and allow them to view solar eruptions or prominences being ejected up into space from the sun's surface. This internal coronograph procedure method has been successfully used with both space and land-based telescopes (e.g., the Hubble space telescope and the Canadian Gemini land-based observatory) to separate the light emitted from orbiting exoplanets from that of their home stars (Fig. 2.2a, b). Since the Hubble space telescope is close to its well-deserved retirement age, NASA scientists have been busy developing the next generation of space telescopes which will commence with the launching of the powerful new James Webb Space telescope sometime between 2014 and 2020 (Fig. 2.3a, b). Some scientists have just recently proposed that an external type of star shade method (Fig. 2.4) may be used in conjunction with the James Webb space telescope (JWST). A separate

Fig. 2.1 Artist's drawing of a special type of planet hunting space telescope that contains an internal device (coronograph) to block the home star's light and allow astronomers to view the light reflected by the exoplanets (image credit: NASA)

maneuverable spacecraft could be launched along with the JWST which could be moved to a location about 100,000 miles in front of the Webb telescope which could then deploy a large sun screen to block the light from stars suspected of hosting planetary systems The JWST could then focus on the light from any nearby exoplanets with minimal interference from the host stars own light.

Fortunately, the advent of powerful high speed digital computers during the second half of the twentieth century has allowed our scientists to come up with another amazing solution to this problem which might eventually work even better than the sun shade idea. What astronomers can now do is to use several spatially separated telescopes such as the Keck Observatory "twin" telescopes in Hawaii (Fig. 2.5) or the newly constructed even larger European Southern Observatory's *Very Large Telescope* (*VLT*) array in southern Chile (Fig. 2.6), which has four large telescopes linked together, to simultaneously gather light from distant celestial objects. The images from each of the separate telescopes would then be sent to a special computer that can digitally "add and subtract" the different individual components (i.e., light waves) from the distant objects' light spectrums to produce a composite image that has much higher optical resolution and clarity than is possible with each individual telescope alone. This new computer assisted multiple telescope technique is referred to by scientists as **interferometry**. In order to eliminate atmospheric effects and obtain much better images than is possible even with the best land-based interferometry systems, our astronomers are now planning on placing **flying formations** of several smaller spatially separated telescopes into space high above the Earth. The separate smaller space telescopes, like their land-based cohorts, will be electronically connected so that the light collected

Fig. 2.2 Photographs showing how coronographs installed inside space telescopes or even land-based telescopes can block a star's light and allow us to directly view the reflected light from orbiting exoplanets. (**a**) Shows a photograph taken by the Hubble space telescope of the actual movement of an exoplanet as it orbits its home star (image credit: NASA Hubble space telescope team), while (**b**) shows a photograph taken by the land-based Gemini Observatory of three exoplanets orbiting their blocked out star (image credit: Christian Marcos et al., Gemini Observatory)

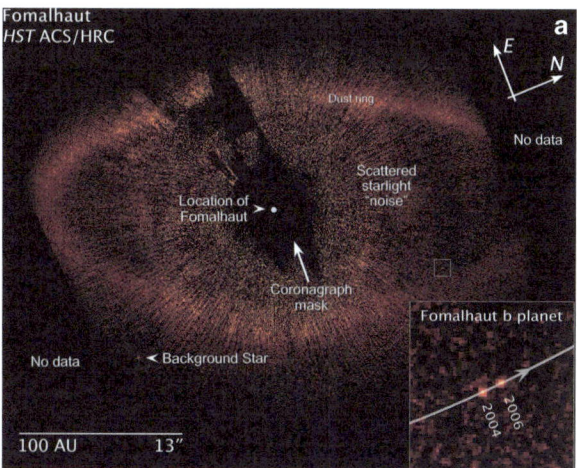

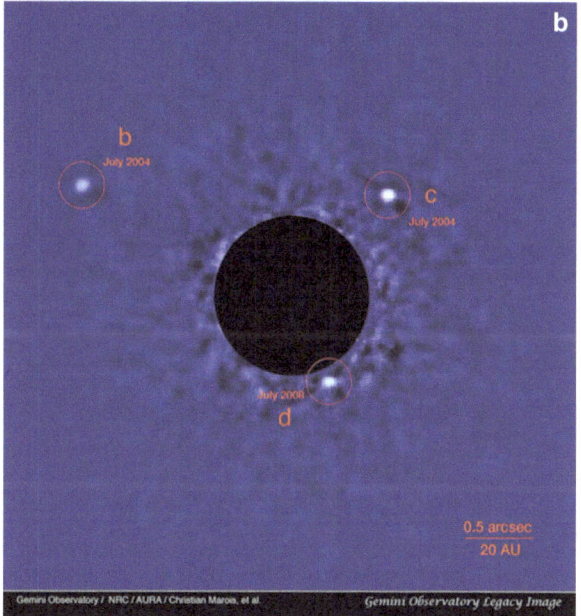

by each of them can be fed into a special computer which will then digitally process and combine the individual images.

Now, how do the computers work their magic to produce this amazing trick? Since the early 1800s scientists have known that light propagates (i.e., moves through space) as waves, like waves on the surface of water. When two light waves meet, they interact with each other. Scientists refer to this interaction as "interference" (Fig. 2.7). If the crests (or troughs) of two waves are coincident (perfectly aligned), they will combine (add) together to produce an amplified wave in what is called constructive interference. However, if the crests of one wave are

Fig. 2.3 Aging causes both humans and space telescopes to eventually retire. After 24+ years of tireless service, having been launched into space in1990, the Hubble telescope is expected to be retired sometime between 2014 and 2020. (**a**) Shows a NASA artist's drawing that compares the aging Hubble space telescope (*left image*) with the new James Webb space telescope which will be launched sometime before 2020 to replace the Hubble telescope. (Note that the telescope mirror of the more powerful Webb telescope is twice the size of the Hubble telescope's mirror.) (**b**) Shows a family portrait of the team of scientists, technicians, and support people proudly standing in front of a life-sized model of their Webb apace telescope science project (image credits: NASA).

completely aligned with the troughs of the other wave, they will totally cancel each other out and the light will disappear. This effect is called destructive interference. Between these two extremes (which scientists refer to as being totally "in-phase" or "180° out-of-phase"), the two waves may be only partially "phase-locked" (aligned) which will permit large variations in the brightness or dimness level of

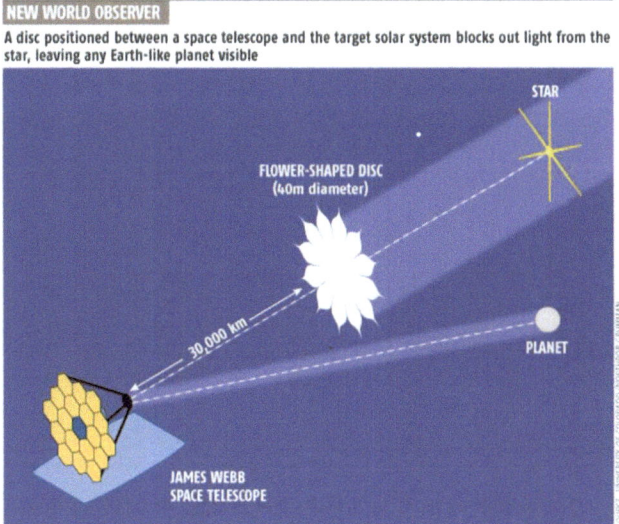

Fig. 2.4 Some scientists have suggested that an external star shade could be launched along with the James Webb Space Telescope and placed approximately 100,000 miles in front of the telescope where it could block out the light from the home stars of any suspected exoplanets and allow the JWST to directly view the planets (image credit: Northrop-Grumman/University of Colorado, Boulder)

Fig. 2.5 Photograph of the Keck twin interferometry telescopes in Hawaii. Two telescopes really can see better than one! (image credit: NASA/Wikipedia Commons)

Fig. 2.6 Shows a photograph taken from a helicopter of the European Southern Observatory's Very Large Telescope Array in southern Chile which consists of four large telescopes that are linked together electronically (image credit: European Southern Observatory/Wikipedia)

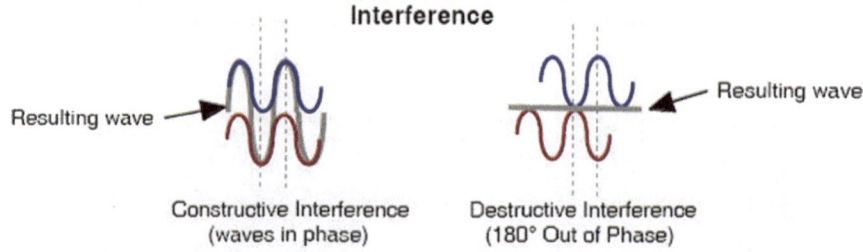

Fig. 2.7 Depicts the two types of interference that occurs when light waves encounter each other in space. When two light waves are in-phase, the light gets brighter (*left drawing*). When the light waves are out-of-phase, they cancel each other and the light disappears (*right drawing*) (image credit: Ricky Leon Murphy, http://astronomyonline.org)

the combined light. Basically, interferometry permits astronomers to *take advantage of the physical phenomena of constructive and destructive interference to digitally separate the extremely dim light being reflected by an exoplanet from the horrendously bright light being emitted by its home star*. Since the home star and the exoplanet are located in slightly different locations in space, and the individual telescopes of the flying formation are also located in different locations, the pathways the different light waves take in traveling from the exoplanet/star pair to the different telescopes will exhibit extremely small differences in both length as well as travel time. Today's computers are able to take advantage of the constructive and destructive interference that results from this situation to mimic the optical resolution capabilities of single much larger telescopes, as depicted in Fig. 2.8. For several years, NASA scientists and engineers have been busily working on

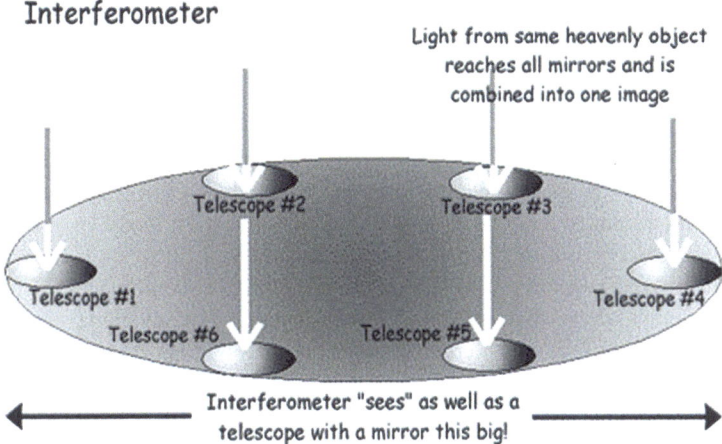

Fig. 2.8 Shows how interferometry telescopes work (image credit: NASA/JPL)

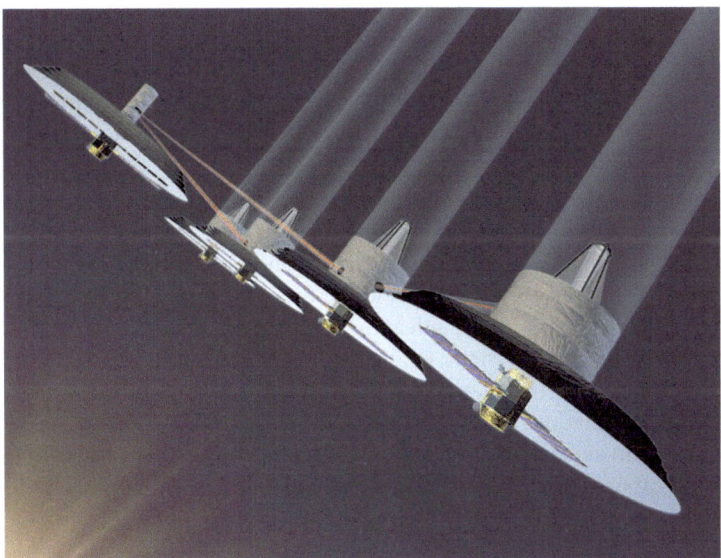

Fig. 2.9 An artist's drawing of what the world's first space telescope system that incorporates the powerful new interferometry technique may look like (image credit: NASA/JPL)

developing flying formations of such individual smaller telescopes which are called **Terrestrial Planet Finder** (**TPF**) systems (Fig. 2.9) that could be placed into space above the Earth (i.e., in heliocentric solar orbits that parallel the Earth's orbit) to assist in detecting and viewing exoplanets.

Although Many ETs May Be "Alien" with Different Chemistries, Science Must Initially Search for Life Forms That Are Dependent on Carbon and Water

Although the newer interferometry telescope systems (both land-based and those placed in space) will greatly increase our ability to detect even the smaller and the more distant exoplanets in our own galaxy, It will likely be many years before we can place telescopes in space or on the moon that will allow us to view distant exoplanets in sufficient detail to see and photograph physical objects or structures on the planets' surfaces that would indicate the presence of advanced technological civilizations. Even the largest engineering structures that ET might place on their planet's surface might be too small to be easily seen with any of our current land-based or space telescopes. Nevertheless, most of our scientists are quite confident that future breakthroughs in telescope technology will eventually solve this seemingly impossible situation. In the meantime, our rocket scientists might be able to develop and construct special robotic space probes carrying sophisticated camera systems or remote sensing technology that could travel at speeds close to that of light (186,000 miles/s). Given the time required for such spacecraft to get up to full speed, it might take 40 or more years for the craft to reach a star system that is located 20 light years away. However, once the photographic session is completed, all the collected data (spectrograms, pictures, etc.) would only require 20 years to be beamed back to Earth. In the next few years, it is quite likely that our astrobiologists will begin to identify some distant twin-Earth type rocky worlds orbiting in their home stars life friendly habitable zones that appear to be warm and wet enough to have a good possibility of supporting carbon-based life similar to ours. Once such planets are identified, our radio astronomers (e.g. SETI) will immediately point their telescopes in the direction of these systems and begin listening for any signs of "deliberate" communication attempts from extraterrestrials such as radio or television transmissions (Ballesteros 2010; Kaufman 2011; Ross 2009; Tarter and Impey 2012; Zuckerman and Hart 1996), or other types of electronic signal "leakage", that might be produced by intelligent inhabitants (Shuch 2011). The SETI scientists will also look for indications of possible visible light transmissions involving pulsing laser beam signals (or signs of nighttime city lights, or even heat related infrared radiation escaping from large ET cities) (Ekers et al. 2003; Valoch 2011; McConnell 2001; Michaud 2007; Mason 2008).

Looking for "Chemical" Rather than "Visual" Signs of Alien Life

However, our scientists will not have to wait until extremely powerful telescopes or extremely fast photographic robotic missions become available in order to begin searching for signs of life on other exoplanets. Since the middle of the nineteenth century, our scientists have had another "trick up their sleeves" that can provide powerful but indirect evidence of the presence of extraterrestrial life. Once we begin launching terrestrial planet finder systems into space that can successfully separate the light from exoplanets from that of their host stars, we can study the reflected light from the exoplanets to determine what kinds of chemical elements may be emitting it. This technique is called *spectroscopy*. Spectroscopy is the use of light to study what physical objects are made up of in terms of their construction, i.e. what kinds of atomic elements they are made of. A device called a spectrometer (or spectroscope) is used to record or measure the spectrum of a specific object or material, i.e., to determine what kind of light it either emits or absorbs. Astronomers can take the light emitted by a star or light from a planet's surface and separate it into its different individual spectral components. Light which is visible to the naked eye is made up of a multitude of different colors. If you pass light through a glass or crystal prism, you will see the light spread out in a series of different colors (Fig. 2.10a). A similar effect occurs when, following a rain shower, sunlight passes through water droplets in the sky to produce a rainbow. However, not all light is visible to the human eye. The human eye is only sensitive to a very narrow portion of what scientists call the total *electromagnetic spectrum* (Fig. 2.10b). All light (i.e., electromagnetic radiation), whether visible to the eye or not, consists of a mixture of different frequencies (colors) that have different wavelengths (some shorter and some longer).

Spectroscopy, therefore, allows light to be collected from a celestial object and then dissected into a multi-colored pattern that shows its spectral content. Since each of the 92+ individual atomic elements that all physical objects in our universe are known to be constructed from have their own unique "signatures" in terms of what color of light they either emit or absorb, astronomers can carefully examine the spectral pattern of the light collected by their telescopes to determine what kinds of specific atoms the celestial objects are made up of. The spectral pattern from a specific source of light will look very similar to that of the bar code that you see in a grocery store on produce labels, except that instead of alternating thin and wide bars that may be light or dark, you will see a series of different colored bands intermixed with a series of black bands. *The different colored bands will indicate which frequencies were emitted, while a black band (no color) will indicate a specific frequency that was absorbed by a celestial object.* For many years, and in many laboratories, scientists have performed extensive studies to determine exactly what spectral patterns are produced by specific kinds of atomic elements (atoms) or molecules (collections of atoms bound together). When astronomers see a particular pattern of spectral lines collected from some celestial body, they can compare this pattern with those collected in our Earth-bound science laboratories to confirm what kinds of atoms or molecules the celestial object is composed of. Figure 2.11a–c shows how different kinds of light spectrums are collected from different celestial objects by astronomers.

Since TPF space telescopes that can totally separate the light emitted by stars from that emitted from the surfaces of at least their larger exoplanets is not yet a reality, our astrobiologists have just recently started using another special form of

a

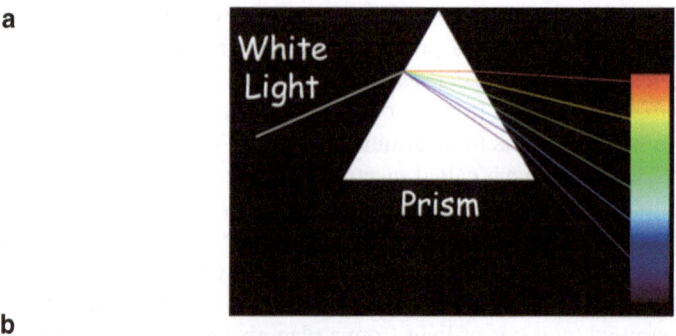

b

THE ELECTROMAGNETIC SPECTRUM

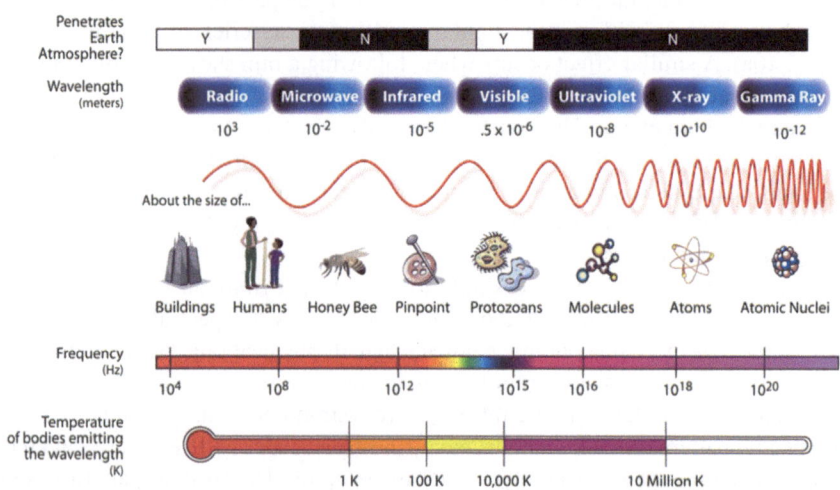

Fig. 2.10 (**a**) When light that is visible to our eyes (i.e., so-called "white" light) is sent through a crystal prism (or droplets of rain in clouds), it becomes separated into a collection of different colors ranging from *red* to *blue* (image credit: NASA). (**b**) Shows that the total range of light energies (i.e., the ***Electromagnetic Spectrum***) is far wider than our eyes can see. The eye evolved while our early ancestors still lived in the oceans. Since much of the total electromagnetic spectrum is blocked by the oceans, eyes evolved specifically to see in the available narrow visible light range. When our ancestors later emerged onto land they still did not need to see beyond the visible range since much of the shorter and longer light wavelengths are also blocked by the atmosphere. Astronomers, however, have found that in order to obtain a more accurate and detailed view of the different celestial objects in space they need to "see" at these normally invisible parts of the total light spectrum (image credit: NASA)

spectroscopic measuring procedure to attempt to determine whether any gases may be present in the atmospheres of distant exoplanets that might indicate the presence of some kind of carbon-based life. This technique is a modified version of the transit method (see Fig. 1.8a) used to detect exoplanets and is called the ***Secondary Transit Method***. This method (Fig. 2.12a) relies on the fact that a transiting exoplanet and its home star sometimes "overlap their images" (when the exoplanet

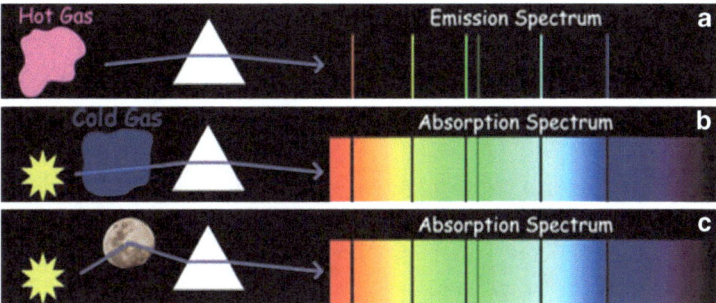

Fig. 2.11 In order to determine what kinds of atomic elements a specific celestial object in space is made of, astronomers typically measure three kinds of spectra. (**a**) Shows the first kind of spectrum. Light from hot objects, such as hot interstellar dust/gas clouds, will display an ***emission spectrum***. These spectrums contain a large number of different kinds of narrow colored bands. The pattern of these bands will tell the scientist what specific chemical ingredients (atomic elements) are present in the cloud that is emitting light energy. (**b**) Shows what is called an ***absorption spectrum***. This kind of spectrum occurs when light from a distant hot object (e.g., a star) passes through a cool or cold interstellar gas or dust cloud. In this case a pattern of black bands will be mixed among the colored bands. The location of the black bands will tell the astronomer what parts of the distant star's light energy is absorbed (i.e., removed) by the chemical elements in the gas or dust cloud and thus reveal the chemical composition of the cloud itself. *Finally, for astronomers specifically interested in determining whether life friendly gases may be present in the atmosphere of an exoplanet*, scientists record a special third kind of spectrum. (**c**) Shows this **"special" absorption spectrum** which involves recording the spectrum of the light that is emitted from the surface and/or atmosphere of the exoplanet itself (image credits: NASA)

is in front of its home star) and, at other times, the exoplanet is behind its home star and not visible. Astronomers can take the "blended" light from the overlapping star and planet and use special computers to subtract this light from that of the star when the planet is behind the star. The resulting "difference" light can then be subjected to spectrographic analysis to determine what kinds of unique chemicals might be present in the planet's atmosphere that are not present in the star's atmosphere. Figure 2.12b shows one of the first successful uses of this Secondary Transit Method. In 2007, Mark Swain and his colleagues from the Jet Propulsion Laboratory in California used the NASA Infrared Telescope in Mauna Kea, Hawaii, to image a Jupiter-size exoplanet that circles a star that is about 63 light years away from the Earth. The scientists used the secondary transit procedure to identify the presence of water vapor, carbon dioxide, and methane in the exoplanet's atmosphere. While these chemicals might suggest the presence of some form of carbon-based life, it is unlikely that this planet is life friendly since it is far too hot (it orbits its home star in a little over 2 days) and is too large to probably support life, at least *as we know it.*

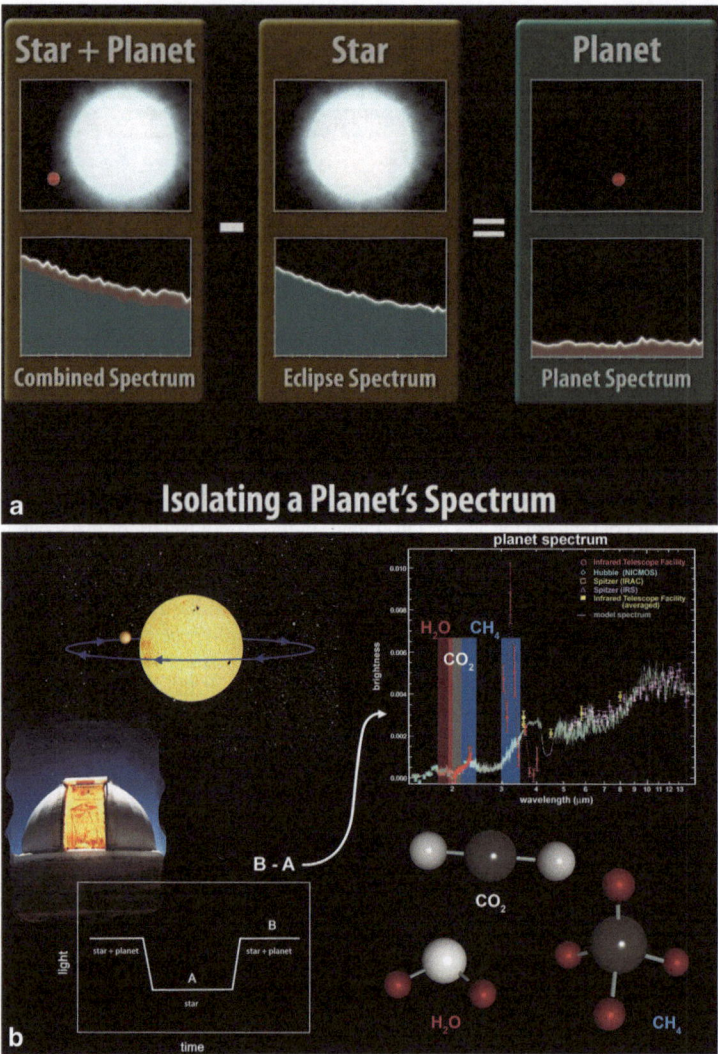

Fig. 2.12 Sometimes an orbiting exoplanet goes behind its home star and at other times it is in front of the star. (**a**) Illustrates the procedure that scientists use to separate the spectrum of exoplanets from the combined spectra of overlapping home stars and exoplanets. (**b**) Summarizes the findings of a group of astronomers that successfully used the Secondary Transit Method technique to isolate the spectra of a distant Jupiter sized exoplanet (image credits: NASA)

Strategies Astrobiologists Will Use to Search for Life on Other Worlds

Therefore, because of economic necessity and scientific conservatism, the search for ET will not initially be focused on finding possible exotic forms of life but on searching for carbon-based life analogous to that found on Earth. Since we currently know nothing about what exotic alien life might look or act like, or be made up of (chemically), our scientists would be totally clueless as to how to even begin such searches (de Vera and Sackbach 2013; Schulze-Makuch 2013; Ward and Bennett 2008). Our best chance for succeeding will be to look specifically for forms of life that we already know exists in the universe. Although our universe is known to be built from at least 92 different kinds of atoms which range from the lightest *hydrogen* atom (which is made up of a single positively-charged proton in the nucleus and one negatively-charged electron that circles the nucleus) to the heaviest *uranium* atom (with 92 protons and 92 electrons), all living things (plants, animals) that we know about on our planet are made up of so-called life friendly organic molecules that are built from a much smaller pool of only six different kinds of atoms (i.e., hydrogen, carbon, oxygen, nitrogen, sulfur, and phosphorus) that were selected from this larger construction pool. Thus, since life is so remarkably rampant and diverse on our own home planet, it would seem entirely reasonable to assume that life based on similar life friendly molecules may have popped up more than once in our vast universe. This means we will need to focus on searching extraterrestrial bodies (planets or moons) where lots of life-sustaining liquid water is thought to exist, and where the different chemical by-products of carbon-based life processes can be easily detected with our current technologies. For example, specific chemicals or gases which, on Earth, are detectable in either the atmosphere or on the surface and are known to be produced by or associated with our form of life will be the top targets on the scientists' search lists. Some examples of such atmospheric gases are oxygen and ozone (which are associated with plant photosynthesis), carbon dioxide (especially when it occurs along with oxygen, could be a metabolic waste product of animals), and methane gas (which on Earth is commonly found in the excrement or poop of barnyard animals and emitted as a metabolic waste product by certain kinds of bacteria). Even more complex chemical substances on a planet's surface itself, such as the chlorophyll content of plant life can be detected from space with light spectroscopy techniques.

In 1990, our scientists actually used this spectrographic technique to confirm that there is indeed evidence that our own home planet hosts some kind of carbon-based life. In that year, NASA launched a special unmanned spacecraft (Galileo mission) to Jupiter that was designed to perform spectroscopic studies of Jupiter's atmosphere to determine what kinds of gases it contained. In order to get the spacecraft to Jupiter, the NASA scientists first had to make it circle the sun, then re-approach the Earth where the Earth's gravity would then act like a sling-shot to "toss" it in the direction of Jupiter. Carl Sagan talked NASA into having Galileo, prior to its being slung in the direction of Jupiter, to perform a spectroscopic study of Earth to see if it could detect any special chemicals that might suggest Earth harbored life. Galileo detected high levels of oxygen and methane in Earth's atmosphere plus signs of chlorophyll on the surface. All three of these chemical ingredients are known to be associated with carbon-based microbes as well as plants and animals. So, first things first—scientists at NASA and ESA have unanimously chosen to give top priority to "following the water" at least for the near future, in hopes of having the best chances of finding life out there. **For at least the immediate future, the absolute _Holy_ Grail of all searches for ET will be, therefore, to focus on finding Earth-size rocky planets that are warm and wet enough to host carbon-based life forms that might be reasonably familiar to our scientists**.

What the first Terrestrial Planet Finder systems will do is to use spectroscopy to look for any signs of the presence of chemicals in the atmospheres or on the surfaces of exoplanets that are *out of equilibrium* (i.e., different) from what would be expected to be present based on known physical and chemical laws of nature as well as the specific type and location (size, chemical composition, distance from parent star) of the exoplanet. If an alien civilization in the Alpha Centauri star system (which is a little over 4 light years away from us) were to build something equivalent to our TPF systems and launch them into space above their home planet, they would probably be able to detect the presence of water vapor, carbon dioxide, methane, nitrogen, and free oxygen in the Earth's atmosphere. All of these gases, in combination, would be a strong clue (but not definitive proof) to the presence of extensive carbon-based life forms on our planet. The presence of free oxygen gas in the atmosphere of a world that does not harbor life would be unexpected since oxygen is notorious for electrically bonding (forming molecules) with other atomic elements and making itself invisible to spectrometry. The reason we have so much free oxygen in our atmosphere is simply because we have extensive plant life that is constantly replacing it. For the first two-thirds or more of the history of Earth, any intelligent ETs living in the Alpha Centauri star system would not have been able to detect our oxygen. It was only following the arrival of cyanobacteria and other forms of plant life which proceeded to rapidly build and inject huge quantities of oxygen gases into our atmosphere that oxygen would have been detectable. Carbon dioxide, while not itself a strong indicator of carbon-based life could, in combination with oxygen suggest the presence of animal life (if not volcanoes or forest fires) that ejected this gas as a metabolic waste product. Large detectable quantities of methane in a planet's atmosphere is also suggestive of life (e.g., especially the larger smelly varieties such as cows, pigs, etc.). Like oxygen, methane cannot survive "out of equilibrium" in an atmosphere without some means of continuously replacing it, most typically via life processes. Small quantities of methane can be, however, sometimes formed via volcanic activity. The recent detection of larger than expected quantities of methane gas in the Martian atmosphere has stirred a debate among scientists as to whether some of the Martian volcanoes might still be active enough to do this, or whether colonies of subsurface methane-producing bacteria (e.g., anaerobic *methanogens*) might be churning it out as a metabolic waste by-product (Fig. 2.13).

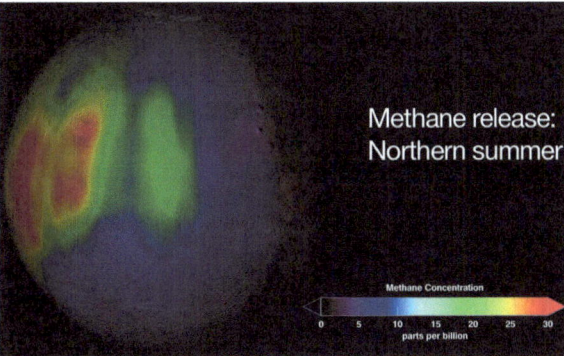

Fig. 2.13 While the surface of Mars is thought to be far too dry today to allow the existence of living organisms, the planet 3 or 4 billion years ago had a thicker atmosphere and a warmer climate which permitted seas, oceans, and running water sources (rivers) to cover large portions of the surface. If any primitive Martian organisms exist today, they would need to have taken up residence in warmer and wetter subsurface regions of the planet. NASA scientists have now found evidence that single-cell lifeforms, similar to those on Earth that exhale methane (CH_4) rather than carbon dioxide (CO_2) as a metabolic waste product may be very plentiful on today's red planet (image credit: NASA)

Scientists Find Evidence that the Chemical "Stuff of Life" May Be More Common in Universe than Previously Believed Possible

In the 1950s many scientists thought that other planetary systems in the universe were extremely rare. Today we know they are quite common. Also, in the 1950s, we thought life was complex and extremely fragile and required just the right kinds of environments in order to develop. The discovery of extremophiles thriving in Earth's most unfriendly environments now suggests life may be tougher than we previously believed possible and might be able to easily pop up elsewhere in the universe.

Life friendly organic molecules found inside meteorites and comets In the latter part of the 20th century, our scientists suddenly started finding evidence of different complex molecules imbedded deep inside meteorites that had fallen to Earth from space.[2] These molecules included amino acids (i.e., the building blocks of life's proteins), sugars, and many other varieties of large molecules (i.e., "macromolecules") that contain what our life scientists now call the *Big Six* life friendly elements, i.e., carbon, hydrogen, oxygen, nitrogen, phosphorus, and sulfur which

---

[2] Meteorites are small rocky or metal objects that are the leftover remnants of the early stages of stellar system formation. These objects (which include the "shooting stars" we see in the nighttime sky) can range in size from grains of sand to several feet or more in diameter. The smaller ones are capable of breaking through the roofs of homes and severely bruising the hips of ladies sleeping on couches, while the larger ones can destroy entire cities.

are required to build our carbon-based form of life. Scientists who specialize in studying the chemical bases for life (i.e., biochemists) refer to any molecule that contains carbon as an **organic molecule**. It is important to note that organic molecules are, by themselves, not "alive" but are the raw ingredients that, when bonded with any of the other members of the "big six" group of atoms in a particular format, and in a particular supportive environmental setting, can become incorporated (at least on Earth) into real "living things"! These large organic molecules are usually composed of carbon atoms bonded to other members of the big six life friendly group in the form of rings or long chains. In the 1950s, almost all scientists believed that outer space was far too hostile (extreme heat and cold, plenty of intense and deadly radiation from stars) to allow complex organic molecules containing large numbers of atoms to be constructed. Therefore, the discovery of such organic molecules imbedded deep inside meteorites was believed to be the result of the meteorites being contaminated (polluted) by the Earth's environment either after landing on Earth or by the scientists themselves who later handled them in the laboratories. However, in a very short time the scientists were able to determine that the organic molecules they were finding inside the meteorites were not the result of contamination but had probably been inside the meteors for billions of years having been placed there when these objects were originally formed from interstellar gas and dust clouds.

And even more exciting is the recent discovery that some extraterrestrial meteorites may, in addition to the basic building blocks for proteins (i.e. amino acids), also contain the building blocks for DNA (Deoxyribonucleic Acids), i.e., those huge molecules that carry the genetic instructions (blue print) for all life on Earth. Dr. Michael Callahan and team members Drs. Jennifer Stern, Daniel Glavin, and James Dworkin from NASA's Goddard's Astrobiology Analytical Laboratory ground up samples of 12 carbon-rich meteorites found in Antarctica and other locations on Earth and used complex chemical and spectrographic techniques to determine that the meteorites contained *adenine* and *guanine* which are two of the four components called *nucleobases* that together form the genetic code that controls how every living organism (microbes, elephants, man, apple trees, tulips, etc.) on our planet is constructed and functions (Fig. 2.14).

At the beginning of the new twenty-first century, thanks to the development of powerful high speed computers, large radio telescopes, space telescopes, and more sensitive tools (spectroscopy) for determining the chemical composition or makeup of distant celestial objects (stars and planets), we also developed the technology to study the chemical makeup of far distant interstellar gas and dust clouds from which new stars and planets are formed. This allowed our scientists to suddenly begin discovering that outer space may be incredibly more life friendly than we dared imagine just a few short years ago. We now have evidence that the atomic elements that are critical for life may actually be chained (bonded) together in outer space to form special kinds of molecules that our life scientists refer to as "prebiotic" (i.e., molecules that are the precursors to actual living or "biotic" groups of macromolecules). Scientists now believe there are special regions or active "factories"

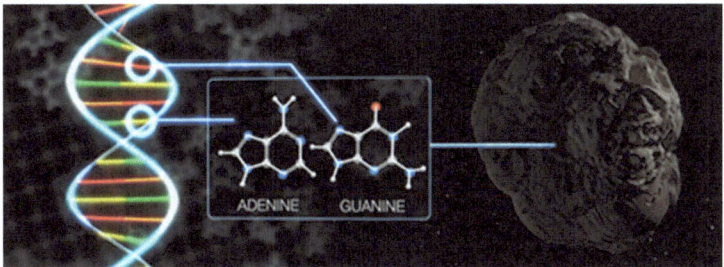

Fig. 2.14 In addition to amino acids, the building blocks for life's proteins, scientists have recently discovered that the chemical building blocks for our DNA genetic materials may also be constructed in interstellar space and then later delivered to new forming planets inside meteors (image credit: NASA's Goddard Space Flight Center/Chris Smith)

located throughout the universe (inside interstellar gas and dust clouds, and especially protoplanetary accretion disks that later produce exoplanets) that are pre-fabricating many of these prebiotic molecules that are then delivered to the surfaces of exoplanets (or the moons of some exoplanets) via meteorites or comets where the final stages of creating actual "biotic" life is then performed (Thomas et al. 2006; Kwok 2012, 2013).

How complex organic molecules are constructed in interstellar dust and gas clouds Before I begin actually describing how and where those complex organic macromolecules that are so critical for life (or at least the kind of life we know about) are constructed, the *diehard educator* in me insists I must make sure my readers understand where our scientists believe such large molecules come from in the first place. If the reader is already familiar with what makes stars shine (i.e., that phenomenon that scientists call "nuclear fusion"), you are free to skip the next few paragraphs of this chapter in which the author will briefly describe how the heavier atomic elements (e.g., carbon, oxygen, and other members of the "big six" life friendly elements plus even heavier elements) are created and made available for the construction of planets and life (Lequeux 2013; Spangenburg and Moser 2003).

As mentioned earlier in the present book, many scientists now believe that, prior to the Big Bang event approximately 13.7 billion years ago, nothing existed in the universe but an unbelievably dense plus extremely hot piece of pure energy that was much smaller than an atom. The first true physical matter created following the Big Bang, however, was not even atoms, but those strange and extremely small "sub-atomic particles" that scientists tell us are themselves the building blocks of atoms Although we now believe there are 36 confirmed kinds of sub-atomic particles, atomic scientists are still searching and will probably discover even more in future years. However, within the first 3 min following the start of the Big Bang expansion (Delsemme 1998), these sub-atomic particles were beginning to link up with each other to form the first hydrogen atoms plus a very small trace of slightly heavier helium and lithium atoms. From that time forward, all the components of our known universe (i.e., matter, energy, space, and even time) were created, and by

380,000 years following the Big Bang all of the hydrogen atoms that we presently have in the universe were created. At first the only physical matter that was created was a huge (a definite understatement) cloud of hydrogen atoms mixed with extremely small traces of slightly heavier helium and lithium atoms. Thus, for all intents and purposes, our young universe shortly after its birth consisted of nothing but a rapidly expanding cloud of incredibly hot hydrogen gas.

Before stars and planets could begin to be formed, however, this huge initial cloud of hot hydrogen gas had to begin breaking up into different regions that contained different amounts of hydrogen gas. Thus, our original gigantic cloud of hydrogen gas that filled the entire universe began to spontaneously break up into a huge number of smaller hydrogen clouds that were surrounded by other adjacent regions that contained far fewer hydrogen atoms. Some of these separate smaller clouds of hydrogen gas then started collapsing (because of their weight, i.e., gravity) and transforming into a much smaller round shape. Starting about 100 million years after the Big Bang, the first stars began to be formed. Many of these stars, which astronomers refer to as our universe's "first generation" stars, were extremely large and massive in comparison to later generations of stars, including today's stars.

As these first giant stars continued to shrink in size, they became increasingly dense and much hotter. Finally, the shrinking ball of hydrogen got so dense and so hot that the nuclei of the hydrogen began to collide with other hydrogen nuclei to produce heavier helium (containing two protons) nuclei. This collision between the nuclei of lighter elements to produce new heavier elements is called "nuclear fusion" by atomic scientists. When two smaller atomic nuclei (e.g. hydrogen) fuse together to produce a heavier nucleus (e.g., helium) a very small amount of the matter is converted to a very large amount of energy (it is this energy from untold numbers of these fusing atoms which is the source of the incredible amounts of heat and light that stars generate). With the onset of nuclear fusion, shining stars are born. Therefore, the first stars created after the Big Bang event contained only hydrogen gas, but when they became dense enough to begin triggering nuclear fusion, new heavier atomic elements began to be formed that contained larger and larger numbers of protons. Eventually, these first generation stars that had gone through the process of creating new heavier elements from fusing hydrogen nuclei together would "die" in a sudden explosion (known as a supernova) and all of their contents (hydrogen plus the newly formed heavier elements) would be scattered into the surrounding space to form what would now be called "gas *and* dust clouds".

This creation of heavier atomic elements by the death of the first generation of hydrogen-filled stars would now set the stage for the development of subsequent generations of stars that now contained both lighter hydrogen plus heavier elements that, when they died, would make it possible for planetary systems to develop. All subsequent generations of star formation, including that of our own sun, would now involve these gas and dust clouds beginning to collapse and getting more dense but instead of forming a dense round ball, many of them would now form a flat spinning disk-like structure (similar to the Frisbees that children play with). The center of this disk-like structure now contained a bulge that would give rise to a new star while the flatter surrounding regions of the disk would give rise to planets. **The reader might want to look again at Fig. 1.2 at the beginning of Chap. 1 which**

shows an excellent artist's drawing of how new stars and planetary systems are formed from these rotating disks of dust and gas clouds.

All stars, nevertheless, eventually die. When stars finally start to run out of sufficient amounts of hydrogen to keep on shining, they briefly turn to the use of helium (which they now have a huge supply of thanks to many eons of producing it in the core of the star) as their new fuel source to produce other slightly heavier elements including carbon, sodium, oxygen, etc. The star's core will begin to take on the appearance of a layered onion with the lightest elements (e.g., carbon, sodium, and oxygen) at the periphery and the heavier elements (e.g. silicon and iron) at the center of the core. However, when the star finishes fusing its lighter elements together (e.g., helium, carbon, etc.) and attempts to start using iron as a new fuel source, it runs into a problem. Unfortunately, iron is far too heavy for the star to use it as a fuel source. When the star tries to fuse iron nuclei together to produce heavier elements, it finds that this process requires more energy than the star is now capable of producing. This causes the outer layers of the star to suddenly collapse onto the core to create a catastrophic supernova explosion.[3] The energy created by this explosion is so intense that, for a very brief time, it does allow the fusion and creation of even heavier atomic elements all the way up to and including uranium. It is, therefore, the explosion of giant and supergiant stars that provides all of the raw materials that are needed to construct everything we see in the universe, including other stars, planets, and even living things. Thus, mother nature is definitely into "recycling"!

However, in order to have the universe that we humans are comfortable with today, we need to have much more than large balls of hydrogen gas that are busy fusing hydrogen to helium in order to shine and produce heat. As diehard carbon-chauvinists, we humans need to be surrounded by a universe that, while still filled mostly with hydrogen (approximately 97 % by volume), also contains a small amount (i.e., 2 or 3 %) of atoms that are heavier than hydrogen. Virtually everything we hold dear in our lives, the ground we walk on, the trees that shade us, and our spouses or significant others that tell us to take the garbage out are composed of heavier atomic elements.

Since scientists have known for some time that virtually all of the atomic elements that are heavier than hydrogen are produced by the fusion of lighter elements inside the incredibly dense and hot centers (cores) of stars during their normal life spans or by the extreme traumatic events associated with supernova or

---

[3] It is important that the reader know that stars that are the size of our own sun or smaller, while making up the majority of all stars in the universe, do not explode when they start running low on their primary supply of fuel, i.e., hydrogen, that is needed to allow them to begin creating their new alternate fuel source of helium and other heavier elements (e.g., carbon, silicon, oxygen). Instead of exploding in a massive supernova event (or an even more intense "hypernova" explosion in the case of much larger supergiant stars), these smaller stars will slowly begin expanding in size to become what astronomer's call "red giant" stars. In the case of our sun, in another 4 or 5 billion years, the sun will expand so much that the outer layer of the new giant star will completely envelop (and burn up) the Earth.

hypernova explosions, we now need to talk about how these different heavier elements are built and put together to form us and the world on which we live. In recent years, our scientists have started discovering that the heavier elements that are ejected into space from dying stars, including helium, carbon, oxygen, nitrogen, phosphorus, and sulfur, plus other heavier elements like iron, gold, and uranium, are mixed together, along with our primordial friend hydrogen, inside huge interstellar gas and dust clouds to form complex collections of grain-like or sometimes fluffy dust particles, many of which are coated with icy materials (Fig. 2.15a, b) These small but complicated structures then become the *interstellar factories* within which complex forms of chemistry can occur (Jastrow and Rampino 2008; Kwok 2013). Various kinds of complex organic (prebiotic) molecules are slowly formed inside these dust grains by a complicated series of different chemical reactions. The

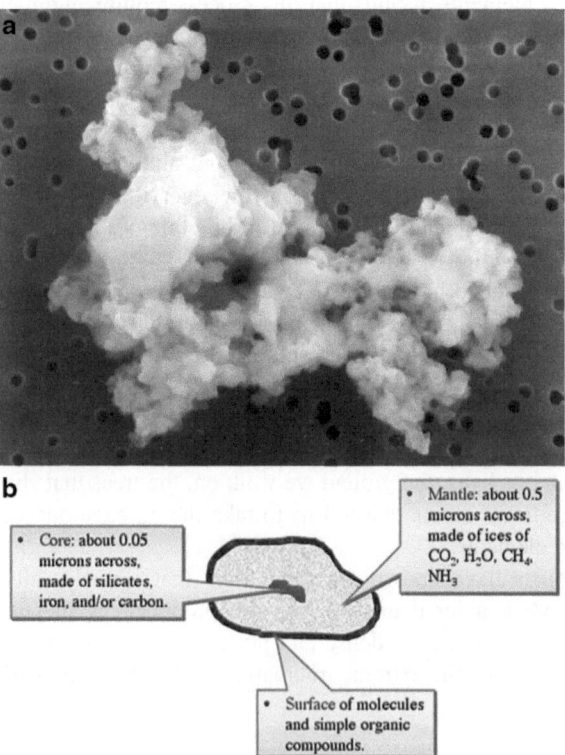

Fig. 2.15 (**a**) Shows a microphotograph of a dust particle that was retrieved by a recent unmanned NASA space mission that gathered and returned dust particles from the tail of a comet. The dust inside the tails of comets (as well as meteors) have been found to contain different kinds of complex organic molecules that originally formed in interstellar gas and dust clouds millions or billions of years earlier (image credit: NASA). (**b**) Shows an artist's drawing of how many interstellar dust particles are constructed. The mantles of many dust grains are composed of carbon dioxide (CO_2), water (H_2O), methane (CH_4), and ammonia (NH_2) (image credit: Centre for Astrophysics & Supercomputing, Swinburne University of Technology)

icy surface of these dust grains both protects the developing molecules from stellar ultraviolet radiation that would normally tear such molecules apart, plus also provides a surface on which atoms and molecules can congregate and chemically interact (bond) with each other to form new and different molecules. Some of the chemical reactions that lead to complex prebiotic molecules only occur in the colder deep interiors of the dust clouds, while others may require the assistance of "just right" (i.e., not too strong nor too weak) doses of heat and radiation from neighboring stars. This "vital dust", as the Nobel prize winning life scientist Christian de Duve calls these organic molecules (de Duve 2002) then finds its way into the interiors of planetary accretion disks where it becomes incorporated into the fabric of growing meteors, asteroids, comets, and other assorted objects that begin raining down on the surfaces of growing planets. As mentioned earlier, analyses of the chemical makeup of meteorites that have landed on Earth have provided evidence that they do contain many varieties of complex organic molecules, including amino acids (building blocks of proteins) and nucleobases (the building blocks of our genes) which carry the detailed instructions for constructing life itself. Dust collected directly from comets by the unmanned NASA spacecraft Stardust probe has also recently been found to contain several varieties of complex organic molecules, including amino acids that are needed to build life (Fig. 2.16).

Fig. 2.16 Artists drawing of an unmanned NASA spacecraft that recently flew through the tail of a comet and collected samples of gas and dust particles to be brought back for chemical analysis. The dust particle shown in Fig. 2.15a was one of the particles that were collected in this mission (image credit: NASA)

Not only are complex organic molecules created in space, scientists can now synthesize them in the laboratory In an amazing new series of experiments, scientists at the NASA Ames Astrochemistry Laboratory and NASA Goddard Space Flight Astrobiology Center plus other laboratories around the world (Cranford 2011; Ehrenfreund et al. 2004; Ward 2005) have been actually replicating in the laboratory the environmental conditions that affect the dust particles that are located inside distant interstellar dust and gas clouds (Fig. 2.17a, b). Various kinds of dust particles (containing various kinds of life critical atomic elements, e.g. hydrogen, carbon, nitrogen, etc.) are placed inside small vacuum chambers that mimic the conditions found inside actual interstellar dust and gas clouds and then either cooled to very low icy temperatures or bathed with ultraviolet light similar to what real dust particles in real interstellar dust clouds would receive from real nearby stars. In the last few years, these scientists have begun to find that many of the complex organic molecules that *occur in living creatures on our planet, or that have been retrieved from comets by NASA spacecraft or found hiding deep inside meteorites that have fallen to Earth,* can now be synthesized in the laboratory!

Therefore, to use a somewhat "far out" analogy, it seems that outer space is not just providing the basic raw ingredients (*building materials*) needed to build life (*e.g. nails, lumber, bricks, mortar, etc.*) but is actually performing some of the critical early construction phases, i.e. pre-fabricating some of the more complex "modular" units (i.e., organic molecules) that, after being delivered to the final building site, i.e. the surface of an exoplanet or its moon, are then put together to form "houses", *or to now drop the analogy*, producing *LIFE*.

Scientists have developed tools to detect and study the chemistry of complex organic molecules in distant interstellar dust clouds In order to be able to examine the chemical composition of the interstellar gas and dust clouds[4] that are located thousands of light years away from us, astronomers had to develop special tools or instruments that could detect incredibly small concentrations of hydrogen atoms that are mixed with even smaller concentrations (i.e., less than 0.1 %) of other elements such as oxygen, carbon, and nitrogen to build organic molecules. And to make matters even worse, the scientists had to study gas/dust clouds that were extremely reluctant to even let us know they are there! The astronomers quickly discovered that the strength or power of the electromagnetic emissions from even the larger gas/dust clouds were unbelievably weak! *In fact, astronomers tell us that the act of letting a small postage stamp fall onto a table top creates a much greater "thud" (i.e., amount of energy or power) than the energy that is emitted by atoms and molecules located inside interstellar gas or dust clouds!* While, with optical telescopes, we can easily detect ("see") the visible light being

---

[4] It is important to note that the concentration levels of even the more dense of these interstellar gas/dust cloud regions are quite diffuse and may contain considerably fewer than 100 atoms per cubic centimeter. In marked contrast, the Earth's atmosphere at sea level today has an average density of approximately 2,500 air molecules per cubic centimeter.

Fig. 2.17 Photographs taken at (**a**) NASA's Jet Propulsion Laboratory (credit: NASA/JPL-CALTECH) and (**b**) the NASA Ames Research Center in California (credit: NASA Ames Astrochemistry Laboratory) show how Earth-bound scientists are able to mimic and reproduce in the laboratory the environmental conditions that occur inside distant interstellar gas/dust clouds. Experiments conducted at JPL and Ames and other laboratories have been successful in synthesizing many kinds of complex organic molecules that are similar to those that are critical to the formation of life

emitted by a gas or dust cloud (especially if the cloud is hot), more often than not the clouds are dark and cold. Still, in order to be able to effectively determine the chemical makeup of such clouds, astronomers do not need to see them but do need to measure the much longer wavelength (and invisible) infrared, microwave, and radio range of electromagnetic energy (see Fig. 2.10b) that are emitted by specific organic molecules in such clouds. Therefore, since the radio waves and infrared heat emitted by distant dark and cold interstellar dust and gas clouds are so weak, scientists had to develop huge telescopes with amazingly large antennae or receiving dishes to have any chance of being able to detect and record such weak signals.[5] Today's best radio telescopes are, therefore, gigantic and very expensive to construct. Figure 2.18a shows a photograph of one such giant radio telescope that radio astronomers in West Virginia have been successfully using in recent years to study organic molecules in distant interstellar clouds, while Fig. 2.18b shows an artist's sketch (diagram) of a group of eight complex organic molecules that the Green Bank radio astronomers have identified in the gas/dust clouds. About 90 % of these interstellar molecules contain the element carbon which is why biochemists and other life scientists label them as being "organic" molecules. Most organic molecules contain at least 6–15 (or more) other types of atoms including hydrogen, nitrogen, oxygen, phosphorus, or sulfur. A very large percentage of these complex organic molecules are similar or identical to many of the organic molecules that plants and animals on our own planet are constructed from. A team of scientists led by Anthony Remijan at the Green Bank Observatory have been able to successfully record the chemical signatures of molecules in a giant gas cloud located 25,000 light years from the Earth that are believed to be the precursors to a key component of DNA and another that may have a role in the formation of the amino acid alanine. The Green Bank scientists believe that the finding of these complex organic molecules in an interstellar gas and dust cloud means that the important building blocks for both DNA and amino acids may be constructed in such interstellar clouds at the same time that new stars and planetary systems are being born which would then be available to "*seed*" newly-forming planets with the chemical precursors for life.

Several groups of astronomers in the United States and other countries have recently started reporting evidence that organic molecules of unexpected complexity exist throughout our Milky Way galaxy. Professor Sun Kwok and Dr. Yong Zhang at the University of Hong Kong, using NASA's Infrared Space Laboratory

---

[5] Because the longer wavelength radio, microwave, and infrared waves originating from far distant interstellar gas and dust clouds are so incredibly weak, the radio astronomers at the Green Bank telescope facility in West Virginia had to push the local government of the surrounding community to actually pass a law to forbid the use of cell phones by their citizens in a several mile radius around the telescope facility. The electronic interference from the average cell phone is so intense that it could easily over power the weak radio signals that the astronomers were trying to collect from the distant dust and gas clouds. So, in a very real sense, thousands of private citizens in the Green Bank area are actively participating as "co-investigators" on this exciting search for life in the universe.

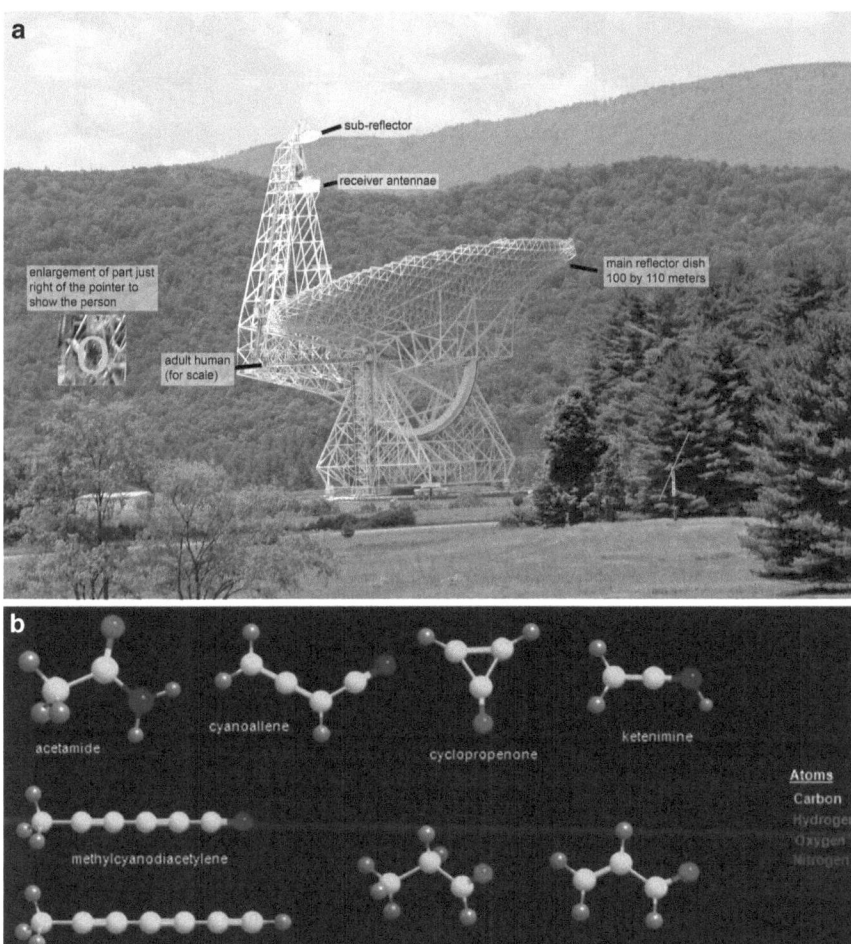

Fig. 2.18 (**a**) Shows a photograph of the National Science Foundation's Robert C. Byrd radio telescope that is located in Green Bank, West Virginia. This incredibly large telescope has so far (as of August, 2006) discovered a total of 141 complex organic molecules located in two typical interstellar dust and gas clouds. (**b**) Shows an artist's schematic model (diagram) of how each of the eight most recently discovered organic molecules are constructed (image credit: Bill Saxton, NRAO)

and the Spitzer Space Telescope have examined the spectra of dust and gas clouds that are located close to distant exploding stars (supernovae) and report that the dust contains a multitude of different kinds of large and complex organic molecules that are very similar to those known to be associated with the development of carbon-based life on Earth (Fig. 2.19). These scientists suggest that such supernova explosions in our galaxy (and other galaxies) eons ago produced shock waves in nearby dust and gas clouds which caused these huge clouds to break apart and

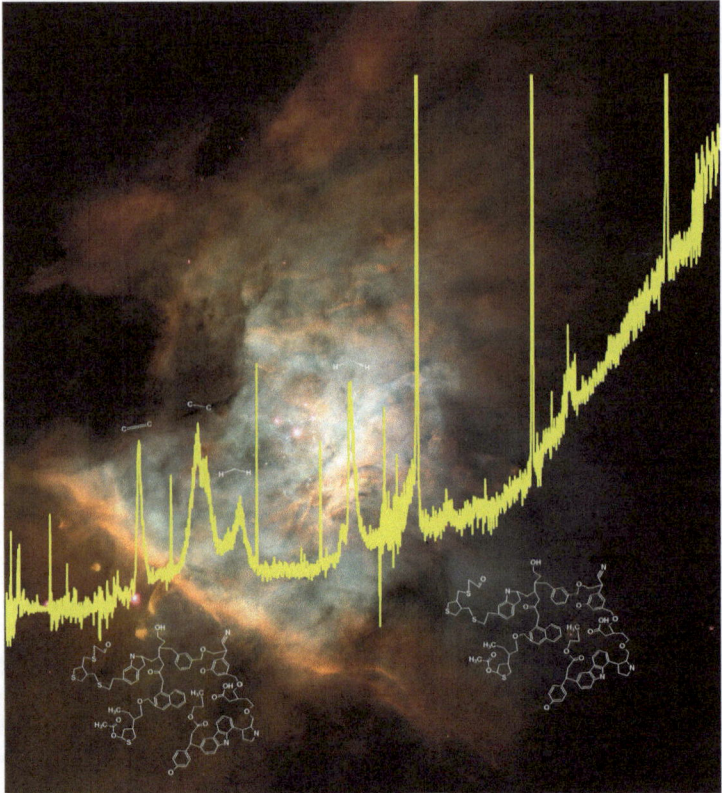

Fig. 2.19 Dust and gas clouds located close to distant exploding supernovas have now been found to contain large complex organic molecules containing carbon and hydrogen that are similar to those needed to build life. The large complex carbon-based molecules which Dr. Sun Kwok found are unusually large and actually resemble coal or petroleum molecules more closely than other simpler forms of carbon molecules (image credit: NASA/C.R. O'Dell and S.K. Wong, Rice University)

collapse into many smaller rotating disks that would eventually give rise to new stars and planetary systems. The dust grains that formed inside these collapsing clouds then began the process of chaining/linking complex organic molecules together and placing them inside meteorites and comets where they would then be delivered via *"local air mail delivery"* to the surfaces of new forming planets to jumpstart life. Many protoplanetary accretion disks throughout our galaxy and even beyond our galaxy may be involved in the manufacturing of complex prebiotic molecules that are needed by any future life friendly planets that may develop in their systems. One of the biggest and most profound questions for our future astrobiologists will be, therefore, determining whether there is a universal formula for developing life (e.g., carbon-based) or whether different kinds of life chemistries (e.g. silicon-based) may develop in different parts of our own Milky Way galaxy or even in far distant galaxies in our vast universe.

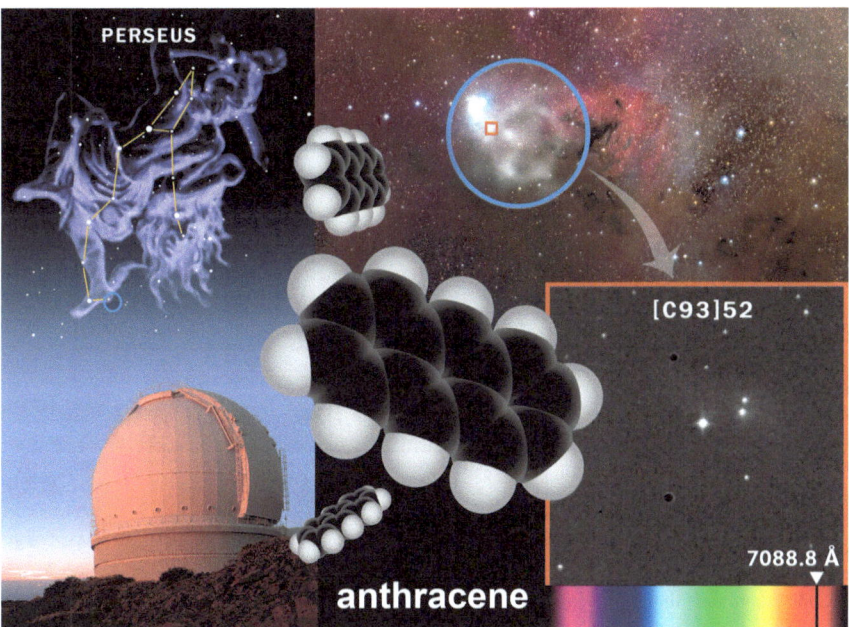

Fig. 2.20 Scientists from the Instituto Astrofisica de Canarias and the University of Texas report finding one of the largest known forms of organic molecules named Anthracene in a distant interstellar dust cloud. This chemical is a common component of coal tar and petroleums (image credit: Gaby Perez and Susana Iglesias-Groth)

Another team of scientists from the Instituto Astrofisica de Canarias and the University of Texas recently found evidence of very large and complex organic molecules in a dense cloud of dust found near a star (Carnis 52, in the constellation Perseus) which is located about 700 light years from the Earth (Fig. 2.20). The molecule is named *anthracene* and is one of the largest known organic molecules in which each molecule contains 10 hydrogen and 14 carbon atoms bonded together. The existence of this large organic molecule in interstellar space definitely refutes science's long held belief that any large organic molecules in space would be immediately torn apart by the intense heat and stellar radiation from close by stars.

Other researchers have also identified the presence of other kinds of complex organic molecules in nearby interstellar dust and gas clouds (e.g., the Orion Nebula) as well as newly forming planetary systems (i.e., protoplanetary accretion disks), including ammonia, methane, silicates, carbon dioxide, acetylene, and hydrogen cyanide (Fig. 2.21a–c). These prebiotic molecules, some of which are the critical gaseous precursors to DNA and proteins, were detected by NASA's Spitzer space telescope inside the protoplanetary dust zone of a young star where rocky planets similar to our Earth are believed to develop. Oxygen molecules, which are critical to some forms of carbon-based life, have also been found in interstellar gas/dust clouds as well as protoplanetary accretion disks (Fig. 2.22).

Fig. 2.21 Ices, water, and other forms of complex organic molecules, including hydrocarbons, methanol, and hydrogen cyanide, which have been proposed as precursors to the building blocks of life, including amino acids and nucleic acids, are being found in many locations in space. (**a**) and (**b**) depict the presence of such molecules in distant protoplanetary disks where new planetary systems are being built (image credit: NASA and Caltech), while (**c**) shows evidence that the planet forming accretion disks of distant sun-like stars may also contain more life friendly chemicals than disks around cool stars. While acetylcholine (C_2H_2) is found in both kinds of disks, sun-like stars contain large quantities of prebiotic or potentially life-forming molecules called hydrogen cyanide (HCN) (image credit: NASA's Spitzer Space Telescope)

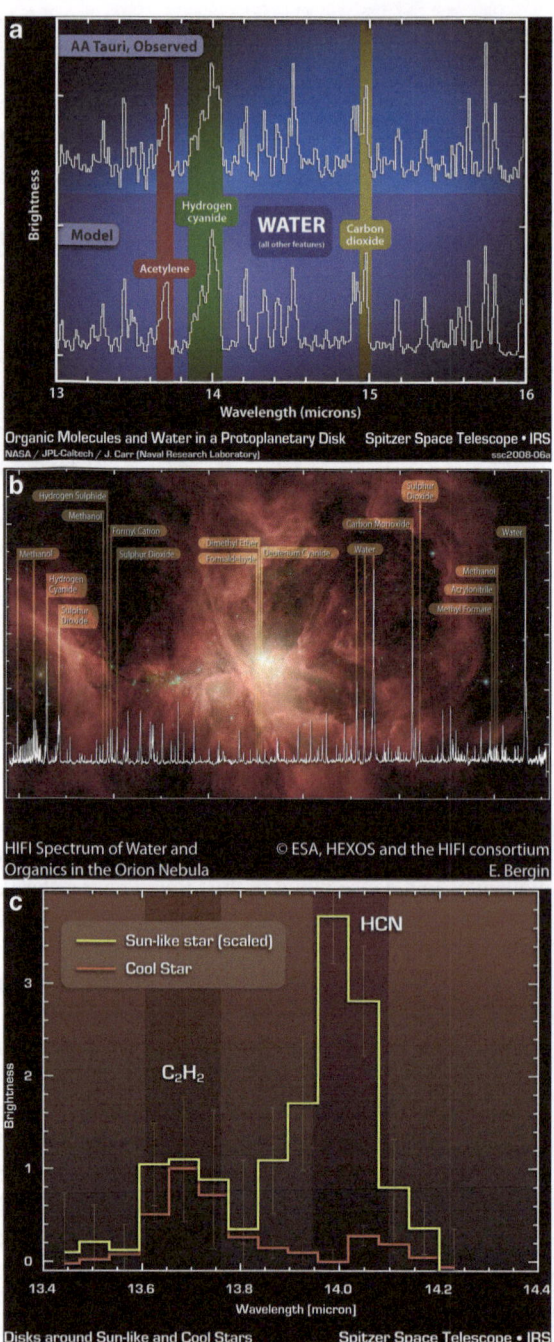

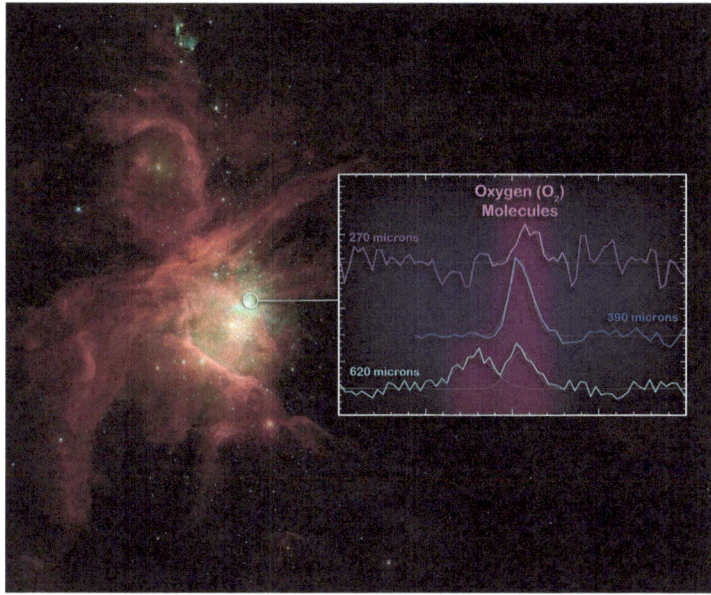

Fig. 2.22 And, of course, some of our scientists have now discovered that the most important chemical molecule, i.e., O_2 oxygen molecules, which are critical for some advanced forms of life comparable to we humans on Earth, are also present in distant interstellar gas and dust clouds (image credit: ESA/NASA/JPL-Caltech)

And, of course, water (H_2O) which most scientists believe is as important for supporting life as is the carbon atom (at least for our form of life), appears to be quite plentiful everywhere in the universe. Our friends at NASA now tell us that water ice is also present everywhere in our own solar system, including even the hottest and coldest environments. Extensive amounts of water ice has been found to be hiding at the bottom of craters on our own moon as well as those located at the north pole of Mercury where they are protected from melting by the shade of surrounding high crater walls.

Scientists have identified life friendly organic molecules in other galaxies in the universe Some scientists are now beginning to expand the search for ET even beyond our own Milky Way galaxy. A team of NASA and European scientists recently used a special form of light spectroscopy to study the light spectra of individual stars in the Andromeda and Triangulum galaxies that are members of the local group of galaxies that includes our own Milky Way galaxy. These two galaxies are located relatively "close" to the Earth at distances of only 2.3 and 3 million light years, respectively (Fig. 2.23a). Another group of astronomers at the University of Massachusetts and Mexico's National Institute of Astrophysics recently used a special type of radio telescope that is setting atop a volcano in a Mexican desert to identify life friendly organic compounds in the Starburst Galaxy (M82) which is located only 12 million light years from the Earth (Fig. 2.23b). The

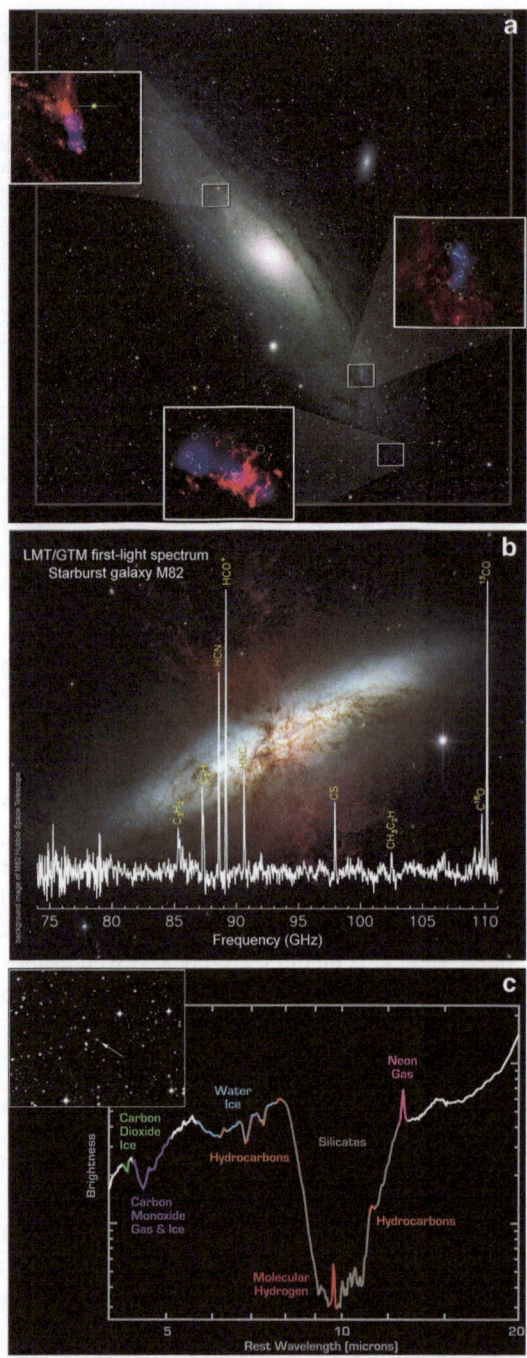

Fig. 2.23 Several types of complex organic molecules that are similar or identical to those needed for life have now been found in other galaxies located beyond our own Milky Way Galaxy. The three *colored insets* in (**a**) are believed to be the spectral light signatures of two types of large

investigators found evidence that certain kinds of large "mystery" organic molecules (that contain at least 20 atoms bonded together) in these stars were absorbing narrow bands of light (which scientists call diffuse interstellar bands or "DIBs" in the visible and near infrared portions of the normal light spectrum. Some of these strange absorption bands are believed to be associated with the presence of large organic molecules.

However, other astronomers (Fig. 2.23c) have found that complex organic molecules are also present in even more distant galaxies. NASA scientists using the Spitzer space telescope, which only sees objects that are emitting infrared light (heat) rather than visible light, have reported the presence of complex organic molecules in another galaxy that is about 1,000 times further away than the Andromeda galaxy at a distance of approximately 3.2 billion light years.

Since the light from all of these distant galaxies took millions and billions of years to reach our telescopes, it would appear that life-friendly organic molecules may have possibly been present in other galaxies in our universe even before life first evolved on our own planet. This remarkable finding suggests that we earthlings may not be the first life forms to develop in the universe. And, as indicated earlier, we now know that our own Milky Way galaxy also hosts many gas and dust clouds that are newborn stellar nurseries where new stars and planetary systems are being formed. The closest such cloud to the Earth and the only one that is visible to the naked eye is the Orion Nebula which is located 1,344 light years away. When our astronomers look through their telescopes at these newly forming planetary systems in the Orion nebula, they are viewing them exactly as they appeared circa 670 A.D. when our human civilization was just entering the early stages of its middle age period and feudal lords lived in castles and our peasant ancestors toiled in the fields to barely eke out a living. *The reader should now look at the Hubble space telescope photograph shown on the cover of the present book for an image of six such protoplanetary systems that are presently being formed in the Orion Nebula.*

Thus, in addition to not "being the first" life forms to arise in our universe, we may also "not be the last"! Our best scientists currently believe that, although the universe we now live in is at least 13.7 billion years old, it probably has many

---

Fig. 2.23 (continued) complex carbon-based molecules located in two of our next door galactic neighbors (the Andromeda and Triangulum galaxies) both of which are local members of the 30 or more galaxies that are located closest to the Milky Way that astronomers refer to as the "local group". (**b**) Astronomers at the University of Massachusetts and Mexico's Instituto Nacional de Astrofisca using the Large Millimetre Telescope which sits on top of a 15,000 ft. high volcano in Mexico have identified several varieties of life friendly organic molecules in the Starburst galaxy M82, which is another member of the Milky Way's local group and is located only 12 million light years from the Earth (image credit: National Institute of Astrophysics, Puebla, Mexico). Finally, (**c**) shows another galaxy located approximately 1,000 times further away (approximately 3 billion light years) that also appears to house substantial amounts of complex organic molecules (image credits: NASA Spitzer Telescope team and NASA Goddard Space Flight Center)

billions of years left before It will either totally cease to exist via some kind of unknown natural catastrophe, or perhaps transition into a different kind of universe or even expand into multiple or even other parallel universes. And, of course, from mankind's own selfish viewpoint, most of us would obviously like to know how many more times our own descendants or other intelligent life forms somewhere out there will be able to witness the birth, evolution, and eventual demise of other planetary systems in other parts of our incredibly vast universe. The life of our universe is indeed the greatest unsolved mystery of all (Deep) Time!

This exciting new research that strongly indicates that different kinds of complex organic molecules may be pre-fabricated in interstellar space has suddenly forced many of our scientists into believing that outer space may be far more life friendly than we could ever have imagined. It now seems that, in addition to the creation of the heavier atomic elements (e.g. elements heavier than hydrogen and helium, such as carbon, nitrogen, iron, etc.) by exploding giant or supergiant stars that are needed to build both planets and life, various "sources" (gas/dust clouds, protoplanetary accretion disks) in interstellar space may also be involved in chaining these elements together to form many of the larger organic molecules that are critical for life and then delivering them to the surface of the Earth and other exoplanets (plus some of their moons). Figure 2.24 shows an excellent drawing by artist Bill Paxton that summarizes how this incredible "interstellar life enhancement" process may work. Thus, while some scientists (including the author) believe it is possible that life based on other kinds of chemistries may exist on some of the so-called "hostile" worlds being identified by the new Kepler space telescope, it is possible that our kind of life (i.e., carbon and water based) may itself be more common in the universe than our most optimistic scientists and even diehard science fiction enthusiasts would have dared believe possible just a few short years ago.

Many scientists, however, believe that there may be two different methods by which life friendly organic molecules can be made available for jumpstarting life. In addition to building such molecules in outer space and then delivering them to the surfaces of planets, some of these complex molecules might also be manufactured by heat-related processes inside hot hydrothermal vents deep in our oceans, or inside hot volcanoes, or even as a result of lightening activity that occur inside thunderstorms in our atmosphere. The bottom part of Fig. 2.25 shows how this other important "in-house" means of building some life critical prebiotic molecules may happen.

What if carbon-based life is not the only kind of life that can occur in the universe? Of course, being the conservative hard-nosed creatures we are, many scientists (including the author) have raised a "cautionary flag" by suggesting that the finding of organic molecules everywhere in space may be an artifact related to our own anthropomorphic biases plus highly selective search techniques. **First** of all, some of us "expect" (and others even hope?) that many (but not all) ETs may, like us, rely on carbon and water-based chemistries. And, **secondly**, all of our sophisticated tools for confirming this are based on technologies (radio and space telescopes, spectrographic and biochemical analysis procedures, etc.) that have

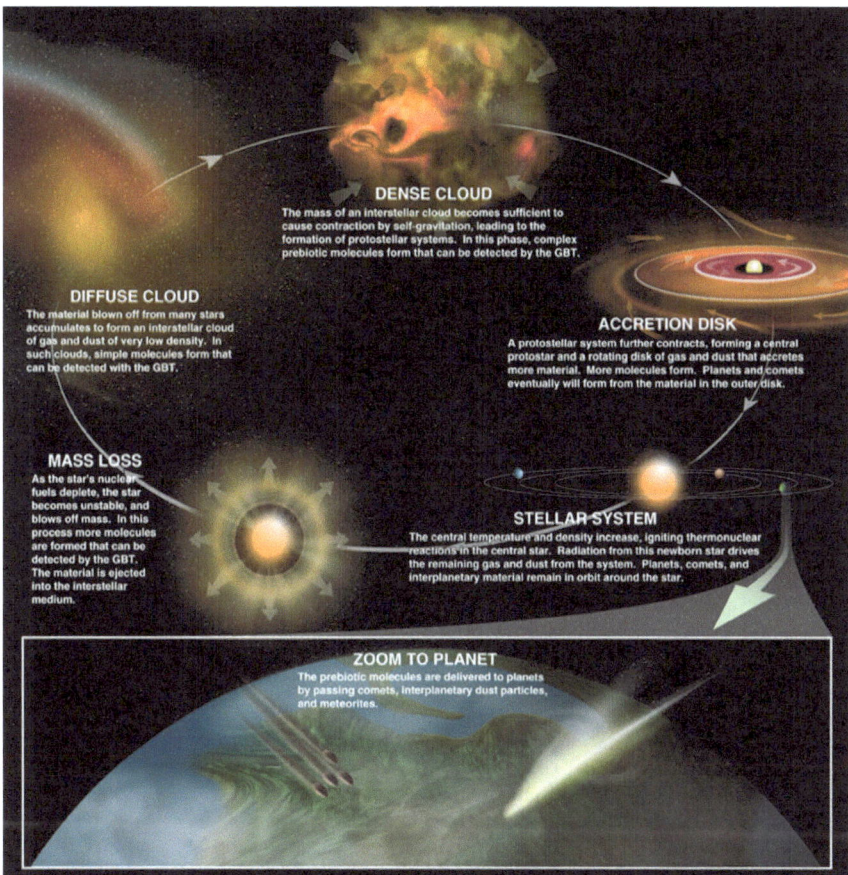

Fig. 2.24 While the transfer (via panspermia) of fully formed living (or dormant) microorganisms from one location to another in the same planetary system might be possible, it is unlikely that this could occur between different planetary systems or other distant locations in the universe. Scientists now believe, however, that it is very possible that many of the complex organic molecules that are needed to build life may be constructed in the protoplanetary disks of new developing planetary systems and then delivered inside comets, meteors, asteroids, or dust grains to the surfaces of their own planets to jumpstart the evolution of life. This excellent artist's drawing depicts the series of events and the actual structures (supernovas, gas and dust clouds, protoplanetary accretion disks, etc.) that are believed to be involved in this complex life building process. Astronomers at the Robert C. Byrd Green Bank Radio Telescope (i.e. the **GBT**) facility in West Virginia (plus similar radio telescope facilities around the world) have been recording the extremely faint radio, microwave, and infrared waves that are emitted by different kinds of complex organic molecules that are constructed when new stars and planetary systems are formed in distant interstellar gas and dust clouds (image credit: Bill Saxton, NSF/AUI/NRAO)

been specifically designed to allow us to detect and identify our own kind of life in our searches. Thus, the finding of carbon-based prebiotic molecules everywhere in space may be largely the result of how and where our scientists are looking and what they are looking for. How many complex molecules based on silicon, or

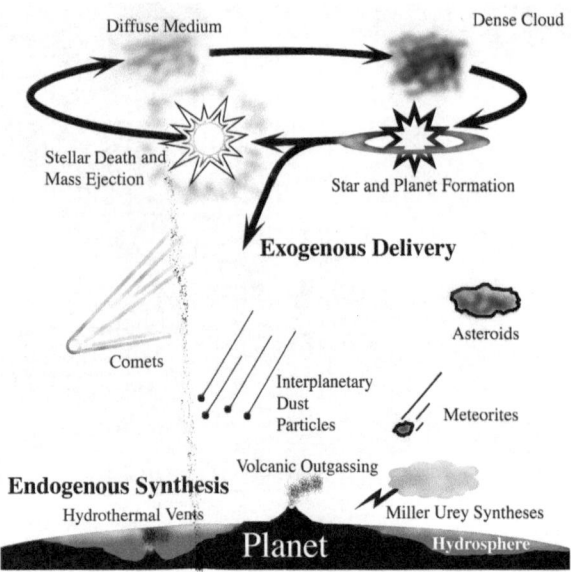

Fig. 2.25 Complex life friendly organic molecules may not always be manufactured "in toto" inside interstellar dust and gas clouds or protoplanetary accretion disks and then later delivered (via **Exogenous Delivery**) to a new planet's surface to jumpstart life. Many scientists believe that some of these organic molecules may be manufactured via **Endogenous Synthesis** in hot hydrothermal vents that are located in the deeper areas of a planet's seas or oceans (if the planet is a water world) or by the mixing of chemicals in the hot interiors of volcanoes, or even as a result of lightening discharges (via Miller-Urey synthesis) in storm clouds which cause simpler molecules in the atmosphere to recombine into more complex organic molecules. This internal or endogenous means of building life friendly molecules is shown in the lower part of this figure. The chemical origins of life on other exoplanets or their moons in the universe may possibly involve one or both of these two complementary but equally important processes. While some amino acids and nucleic acids may be built in distant interstellar dust clouds and then delivered directly to a planet's surface inside comets, asteroids, or meteorites, other types of amino acids or other prebiotic molecules might be built from scratch using special chemicals and the assistance of energetic heat sources inside volcanoes or hydrothermal vents (image credit: NASA Ames Astrobiology Research Institute)

phosphorus, or boron, have we found in interstellar space? And, how hard have we looked? Most scientists believe that complex molecules based on silicon are much less stable than the carbon-based varieties and, therefore, less likely to provide a viable basis for the chemistry of life (ergo, "life as we know it"). Such silicon-based molecules do appear to be less stable (and, therefore, less life friendly) on Earth, but how about other extreme environments in space? Have we performed adequate research to verify this critical difference between carbon and silicon, or have we allowed our personal biases to restrict our investigations and constrain our interpretations of the data? After all, our scientists have been accused by more than a few members of the popular news media of being carbon-chauvinists! Whether ET is boron-, phosphorus-, silicon-, or carbon-based would likely have very little to do with determining whether their outer morphological appearance is in any way similar to that of us long-legged, lanky, and big-headed humans. Some of those giant ugly grasshopper-like creatures that the science fiction people have conjured up as totally alien might not look anything like us but could still be our "first cousins" in terms of their chemistry. On the other hand, some silicon, boron, or phosphorus-based creatures might, if they evolved in virtually identical environments to ours, develop remarkably similar external morphological characteristics to ours since evolution tends to be conservative in using the easiest solutions to allow living creatures to adjust to similar environments. Mammals that chose to return to the oceans (e.g., whales) traded in their legs for fins in order to swim like fish. And, of course, eyes and wings have been reinvented over and over again to allow different animal species to see and fly.

Finally, before ending this chapter, I would like to discuss some very strong personal biases that I, along with probably many other "owner's of human brains" have with regards to what life is and what kinds of life may be out there in our vast universe. Most human brains work virtually the same, having been molded and shaped by a common biological and societal history and a common physical/ chemical environment that we all share. In this regard, human brains worldwide are virtual clones of each other. It is no surprise, therefore, that most of us (scientists as well as non-scientists) tend to be close-minded when it comes to considering alternative possibilities to the things we think we know or have learned beginning at a very early age. And, whether they admit it or not, this statement is equally true for the science fiction writers as it is for all the rest of us.

The author freely admits that it is very difficult for him to conceive of alternative types of life forms that could exist elsewhere in the universe that might be markedly different than those that we are familiar with here on Earth. Although living species on our planet come in an amazing variety of different physical sizes, colors, shapes, as well as behavioral characteristics, etc., they are all based on the same chemistries (i.e., the carbon atom and its five "buddies", i.e., hydrogen, oxygen, nitrogen, phosphorus, and sulfur) and all must have easy access to water. Elephants, amoebae, and palm trees may look radically different but they are still our first cousins in terms of their chemical makeup.

At the present time, all of our astrobiologists and other life scientists (plus our science fiction cohorts) would have to admit that, as far as we now know, the only kind of life that exists in the universe is carbon-based and water-dependent, and we are it! The question of whether this is the only kind of life that can possibly exist is, at present, completely unknown. Since we presently do not know whether alternative kinds of life might exist, it makes total sense for our space scientists to focus their initial searches on finding twins of our planet Earth, or reasonable possible alternatives thereof (i.e., not too large or small, too hot or cold, with an oxygen atmosphere, and adequate quantities of easily accessible carbon and water, *either now or sometime in the past*). In the far distant future, if our space faring descendants do discover that all life everywhere is carbon-based and dependent on water, I personally would be surprised but not too surprised. While it is possible that the inherent chemical and physical features of the carbon atom might make it the leading candidate for supporting life in vast regions of the universe, it is very possible that environmental conditions somewhere out there may have allowed silicon (or boron or phosphorus) to be the primary building block, and water (while it, like carbon, appears to be everywhere) might have lost out somewhere else to liquid methane, or other types of solvents, as the primary liquid facilitator of life. Of course, as is definitely the case on Earth, the carbon/water life combo in other parts of the universe most likely would assume an incredible variety of different physical or morphological formats, some of which (like they do on Earth) would look totally different from what we believe a living creature should look like. And, as for communicating with ET, our scientists may discover that, while many of these strange looking creatures may live in the same universe as we do, they may think, behave, and interact with their environments in a manner that is totally alien to us. When it comes to having a mental or psychological makeup that is similar to ours, we may discover that we have more "in common" with our own family pets (e.g., cats or dogs; or even parakeets) then we do with many ETs. Thus, the question of whether we are alone (if not physically, then possibly psychologically) in the universe may turn out to be more difficult to answer than we now believe.

Finally, it is important to note that all of the current techniques SETI is using to try to identify and locate intelligent life in the universe are based on *"what we think we know"* about what life is and whether or not it might even want to communicate with us (Ballesteros 2010; Tarter 2008; Tarter and Impey 2012). Looking for radio or pulsing laser light signals coming from space is, of course, worthwhile, since some ETs could be similar to our human species and might choose this mode of communication, but many other ETs may rely on techniques that would be totally different from what we might expect or even consider possible.

Chapter 3
If Other Kinds of Nervous Systems Exist in the Universe, How Do We Locate and Identify Them?

The idea that scientists could take the one and only example of a biological control system that developed on planet Earth, which we call a "nervous system", and compare it with similar or analogous systems that we do not yet know even exist on other worlds would seem to be a totally impossible task. Yet, those of us who call ourselves astrobiologists are now finding that our new science is quickly approaching a critical point when we will need to have this information, or at least have developed a sound strategy for obtaining it when we get the opportunity which, hopefully, will not be too far in the future. With the recent finding that life on our planet is far older, far more diverse, and far more resistant to destruction than previously believed (i.e., the discovery of extremophile forms of life everywhere), along with the sudden and rapid surge of discoveries of possible homes (exoplanets or their moons) for alien life everywhere in space, many of our space and life scientists are now beginning to feel the urgent need to begin making preparations for investigating what may be the next big breakthrough in astrobiology related to determining not only what life "is or is not" but also how it functions and interacts with its own surrounding environments. Our astrobiologists need to now start thinking about how ETs, if they exist, might interact not only with their own environments but perhaps also with us humans who might be about to become "their" new fellow inhabitants of "our" universe.

So, how do we scientists go about possibly comparing something we know a great deal about (i.e., biological brains or nervous systems on our planet) with "something" (i.e., ET nervous systems) that we know absolutely nothing about? I would like to believe that, since my professional background is in the brain sciences, that I might have some kind of advantage in taking on this seemingly impossible task. Those few people like me, who are hooked on both astronomy and neuroscience, might possibly have an even greater advantage in doing this since their dual knowledge of astronomy/space science as well as the brain sciences might possibly assist them in focusing more specifically on any possible interconnections between their two major scientific interests. Since this author, along with everyone else on Earth, presently has absolutely no knowledge of any

© Springer International Publishing Switzerland 2015
J.L. Cranford, *Astrobiological Neurosystems*, Astronomers' Universe,
DOI 10.1007/978-3-319-10419-5_3

physiological, biological, or neurological "stuff" or "things" that are directly linked or tied to the broad field of astronomy, what advantage could someone like me have in being knowledgeable in the Earth-bound brain sciences? If I know a considerable amount about one of the two subject areas and absolutely nothing about the other, how can I make any reasonable comparisons? It is even possible that my knowledge of our Earth-bound form of brain science might inflict me with totally erroneous biases with respect to any such comparisons, especially if the neurobehavioral lifestyle of ETs turns out to be as complex and diverse elsewhere in the universe as many of us now believe that life itself may be. However, my knowledge of how one type of nervous system works on our own home planet might put me in the unique position of being ready to "leap in" and take advantage of any real solid neurological information that future astronauts might stumble upon with respect to any kind of neural processes or structures they might run across in their encounters with ETs. The fact that this author is trained and knowledgeable in the neurosciences plus also interested and knowledgeable in astrobiology, could possibly allow me, or persons with similar dual science backgrounds, to make some kind of positive contribution to these two fields of science. In the following three chapters, I will attempt to set the stage for any future discoveries of actual neural type systems on other worlds in the universe by describing, using as much layman's terminology as possible, how such systems evolved and developed on our own home planet. I sincerely believe it is now time for us to begin exposing our astrobiology graduate students to such psychological or neuroscience information to allow them to be ready when the first communicating ETs might come knocking on our doors.

Now that most scientists believe that living creatures may exist elsewhere in our own solar system or on far distant exoplanets that are capable of at least responding to, if not thinking or interacting with their environments, I will now put on my brain science hat and begin a serious discussion of this other complex topic of the present book. In order to understand how intelligent nervous systems might evolve on other worlds in the universe, we must look for samples of such systems and thoroughly study them to find out how they develop (evolve) and function. Unfortunately, the only example of a nervous system we know of is the one that evolved on our own planet. This single example of a nervous system "as we know it" must be our starting point in attempting to speculate on how other nervous systems might develop elsewhere in the universe. Fortunately, scientists on Earth have, for many years, been intently studying how their own and the brains of their Earth-bound relatives developed. Since I was fortunate enough to be a productive member of this "brain scientist" club for over 40 years, I will now attempt to describe how all of this happened on our own planet and how it might happen elsewhere in our incredibly vast universe.[1] In this chapter, I will undertake two major tasks. First, I

---

[1] However, before going any further in this bizarre attempt on my part to describe how I (even as the lifelong professional neuroscientist that I am) think nervous systems might be able to evolve on other worlds, I must unequivocally and very strongly emphasize that while I know a considerable

will attempt to list and describe at least the most basic and life critical functions that any Earth-bound nervous systems, and possibly also extraterrestrial or alien nervous systems elsewhere, must perform in order to assist living creatures in their daily lives. And, following this, I will turn to a description of the general anatomical and physiological means by which our own Earth-bound nervous systems are constructed and work. I promise to do everything in my power to keep this latter anatomy/physiology lesson as simple as possible. As a lifelong neuroscientist, I can personally attest to the fact that the anatomy and physiology of the human brain is extremely complex, and definitely rivals the extreme complexity of how the universe itself is constructed and works. When astronomers assert that there are at least as many stars in the universe as all the grains of sand on all the Earth's beaches, the brain scientists quickly retort back "*Ditto for the number of individual nerve cells in the human brain!*"

Because of the extreme complexity of even the simplest Earth-bound nervous systems and their functions, the reader needs to be forewarned that this discussion will, of necessity, be restricted to a description of only the most basic fundamental aspects of the function and anatomy/physiology of such systems. Rather than describing how a human nervous system might allow its owner to compose a musical symphony or win a gold medal at an Olympic figure skating competition, I will actually have to place the bar much lower in order to not totally lose the reader (and myself) in an absolute morass of extremely complex information and scientific data. I might be tempted to try to describe, for example, how the pea-sized brains of salamanders assist them in their daily activities, but to do so would require me to write more information than could be placed in several book volumes or comfortably stored on my computer's large hard drive, and my small cabbage sized brain would simply not be up to this task (and some of my readers would probably toss this book in the trash). Still, the radical simplification that I am proposing for the present book should work quite well for purposes of speculating about the functional aspects of extraterrestrial nervous systems since **the nervous systems of all animal species on Earth function basically the same and are constructed in the same basic manner**. Whether there are any universal laws or rules for how nervous systems are constructed and work is, at present, totally unknown. It would be "great" if there were, but our future astronaut-scientists must be prepared for the very real possibility that such systems may come in an amazing variety of different "strange" forms.

---

amount about how brains developed on Earth, I, like the reader and everybody else on Earth, know absolutely nothing about how, or even whether, nervous systems exist elsewhere or could even develop elsewhere in our obviously vast and hostile universe. However, I strongly believe that with all the incredible discoveries of extremophiles and exoplanets that our scientists have made in the past few years, the possibility that we are not alone in the universe is now definitely growing. It is now time for some astronomy student out there to team up with someone with knowledge of how brains develop and work on Earth, and take the lead in trying to merge our rapidly exploding fields of both astrobiology and brain science. Perhaps it Is time to develop doctoral level training programs (or at least coursework) in "astrobiological neurosystems", "astroneurobiology" or "astroneuroscience", or some such "moniker" that we could impress our colleagues with.

Most scientists now believe all life on Earth evolved from the same common ancestor which was quite possibly an extremely small bacterial size creature that first popped up near some hot hydrothermal vent system deep in one of the first oceans. All life on Earth, whether bacteria, plants, or brainy animals, share the same genesis (origin) that is based on water and the carbon atom. Life forms on another world might also be carbon-based life and share a similar genesis to ours, or be the result of a totally different genesis because of drastic differences in the environment in which they developed. Instead of being based on the carbon atom and water, these alien creatures might use another atomic element such as silicon for constructing their bodies and/or a different solvent such as liquid methane to facilitate their biological chemical reactions.

How Nervous Systems on Our Planet Facilitate Their Owner's Lives

Strictly from the point of view of evolutionary biology, the "nervous system" (or its ET equivalent) of any living organism carries on only two basic but highly critical functions. The primary function or job of the nervous system of any animal species is to exert centralized control over the functions of all of the other organs or biological systems of its owner's body. The second job is to provide a means of allowing organisms to interact with their surrounding environments so they will be able to reproduce and avoid their own destruction. In all animals on Earth the nervous system is composed of a number of separate structures.

*Before going any further in discussing how nervous systems work on Earth or might work on other worlds in the universe, I must briefly pause and promise the reader that, in spite of being a professional neuroscientist, my description of the neuroanatomy, neurophysiology, and other technical aspects of the field of brain science will **NOT BE OVERLY COMPLEX**. In contrast to many other fields of science, the fields of astronomy and space science are fortunate in the fact that many authors, because of the inherent popularity of these two fields of science, have been able to write excellent books that are simple and non-technical enough that general non-science trained readers can understand. The present author has made every attempt to write the present book with that objective in mind. Especially in the brain science and evolution chapters (Chaps. 3 and 4) and also the computer-related Chap. 5), I have only suggested a small number of books by other authors that I sincerely believe most interested lay readers will be comfortable with. I do hope I have succeeded in this simplification task, and would like to invite comments and suggestions from my readers. So, let us begin.*

Figure 3.1a shows the different anatomical (structural) parts of the normal human nervous system (Amthor 2012; Dubin 2002; Liebman 1986). The largest structural part of man's nervous system, which is known as the **brain**, is located in the front portion of the body (typically referred to as the head or skull), usually close

Fig. 3.1 (**a**) Shows a medical artist's drawing depicting the location of the *brain* and *spinal cord* components of the human *central nervous system* as well as the location of the *nerve fibers* that enter and exit the spinal cord to form the *peripheral nervous system* (image credit: U.S. National Library of Medicine/www.nlm.nih. gov/). (**b**) Shows the functional (i.e., hearing, touch, smell, etc.) areas of the human cortex (image credit: Purves et al., *Life: The Science of Biology, 4th ed., Sinauer Associates, 1998*)

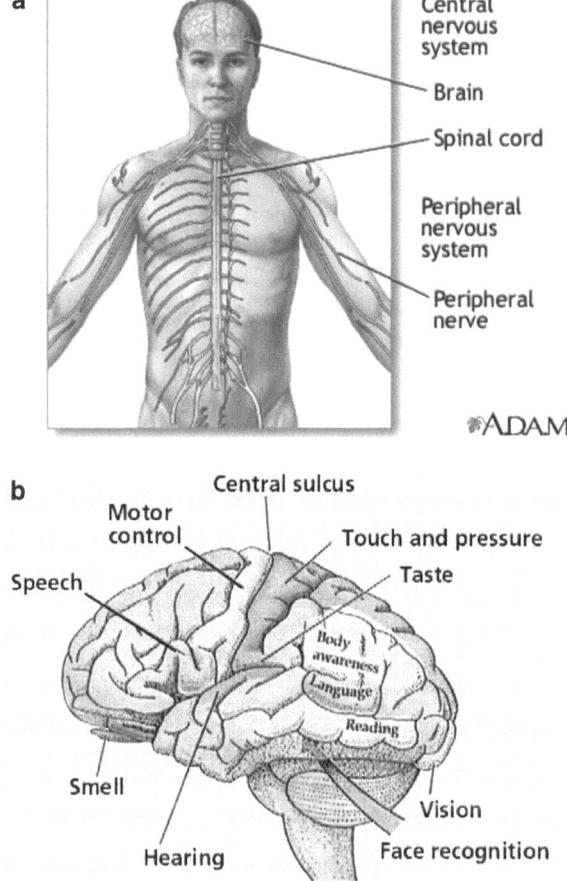

to the sensory organs that control vision, hearing, balance, taste, and smell. The parts of the brain that receives and processes the neural information from these sensory organs is shown in Fig. 3.1b. In addition to the brain, our nervous systems consist of a **spinal cord** that is located inside a protective bony structure (known as the spinal or vertebral column) that extends from the head of the animal to its rear end (buttocks, anal region). Together, the brain and spinal cord are known as the **central nervous system**. At all points along the entire length of the spinal cord, long **nerve fibers** (i.e., **axons** and **dendrites**) travel to and from various structures and organs in the body. These long nerve fibers transmit electrical nerve impulses to and from the spinal cord and the brain and are referred to collectively as the **peripheral nervous system** (**PNS**). Their job is to transmit sensory information on the status of the organs and other peripheral structures (e.g., muscles, joints), as well as what is happening in its owner's surrounding environment, to the brain and in turn allow the brain to transmit instructions as to what to do (e.g., flex certain muscles, release specific chemicals to make a particular gland or organ to do its work, as well as allowing its owner to interact with its external environment, etc.).

Therefore, to use a somewhat strange analogy that the author occasionally used in his brain science lectures with his students, it seems that the brain, spinal cord, and peripheral nerves all work together as a complex inter-communication system that is not unlike a *busy highway freeway system* (Al-Chalabi et al. 2006; Dubin 2002; O'Shea 2006; Jerome 2003). The human or animal brain takes information it receives from the outside world or its owner's body (e.g., eyes and ears located in the head, or sensory receptors located in the tissues, organs, and muscles elsewhere in the body) and transmits this information over the *freeway system aka spinal cord* to the brain. The brain then determines (based on automatic reflexes or stored memories) what is "happening" and "what needs to be done" and sends instructions over the spinal cord to its owner's organs or body structures where "things happen".

It is obvious to almost everyone on Earth that the more advanced the animal species, the more their nervous systems do for them. On Earth, nervous systems, and especially the brain, expand rapidly in relative size and complexity as we ascend the evolutionary ladder from primitive animals to man (Fig. 3.2a, b). Figure 3.2a contrasts the two extremes of the range of nervous system complexity found on our planet, from primitive insects, worms, and squids to us humans At the other extreme, Fig. 3.2b shows the wide range of size/complexity of the mammalian line of evolution from rats to advanced dolphins and humans. It would be extremely difficult for the author to think of, or even list, all of the different functions that even the simplest of Earth's nervous systems can perform, and totally impossible to list those inherent to the most advanced nervous systems, such as our own.[2] Animal scientists now tell us that even the most primitive single-cell organisms (microbes) can "learn" how to locate food and avoid noxious stimuli in their environments. And, of course, man's brain can do much more, albeit not always effectively or accurately. While rats can learn how to effectively navigate through a complex maze to more quickly locate food, man can build and launch geological positioning satellites (GPS) to quickly tell him/her where the nearest fast food restaurant is located on Earth. More advanced nervous systems do more and more complex things to assist their owner's lifestyle than do primitive nervous systems.

---

[2] The author would like to state totally upfront that he believes the single greatest impediment to our neuroscientists discovering exactly what kinds of functions different animal species are capable of is that our social history, due to our being predatory and emotional beings has totally ingrained many of us with the idea that, as a species, man ranks directly below God and the angels in terms of our mental capabilities. Mankind's pervasive **anthropocentric** concept (the idea that man is unique in being "god-like" in having large varieties of complex mental capabilities, while "lower" animals have very few such higher level mental capabilities) has plagued the brain science literature for centuries and caused many of us to believe that consciousness, problem solving skills, language, and many other complex mental activities are quite common in man but totally (or near totally) absent in most lower animal species. Man "thinks" and "solves problems" while lower animals rely on inherited or ingrained built in or prewired instinctive mechanisms. When our astrobiologists, in the next few centuries, begin finding complex living creatures on other worlds, our neuroscientists must dump this anthropocentric constraint to their professional thinking, or they will fail miserably in their work.

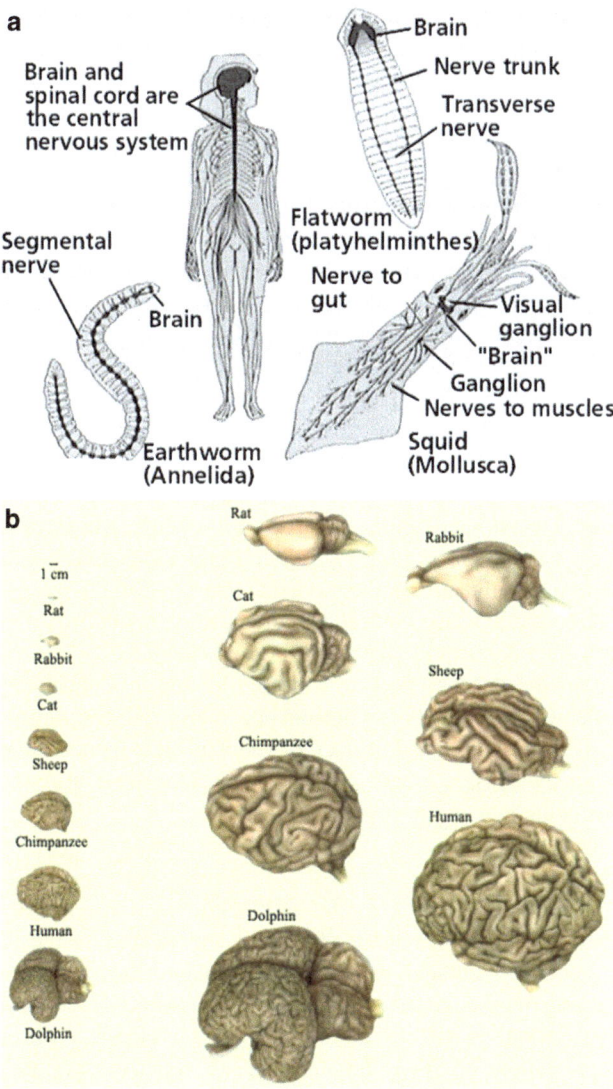

Fig. 3.2 (a) Shows a drawing that contrasts the human nervous system to those of several primitive animal species. Even worms have central nervous systems (brains) and peripheral nervous systems (e.g., segmental nerves, transverse nerves, nerves to muscles, etc.) (image credit: Purves et al., *Life: The Science of Biology*, 4th ed., Figure 38.2, page 850, Sinauer Associates, 1998). (b) Shows how the neocortical part of the cerebral cortex has expanded in both size and complexity from more primitive mammals to humans (image credit: www.thebrain.mcgill.ca)

Probably the most important function (life critical) of any nervous system is to physically control and coordinate the multitude of chemical, physiological, and behavioral processes that are needed to keep its owner alive and well in a hostile environment that is constantly threatening to destroy it (Asimov 1994; Kalat 2008; Nolte 1988; Sagan 1986; Slaughter 2002). The nervous system must constantly make sure the internal physiological and chemical state (environment) of the organism remains different (i.e., out of equilibrium) from that of the outside world. The nervous system must control and regulate all of the many different chemical and physiological processes involved in, to name just a few functions, breathing, feeding, growth, reproduction, etc. These life-supporting functions are what we normally refer to as "bodily functions" as opposed to "mental functions". Higher organisms, such as humans, are apparently aware (i.e., conscious) of many of these bodily functions, but definitely not all. To protect themselves from the hostile outside world, all organisms, from single-cell microbes (such as bacteria or amoebae) to multi-cellular creatures (e.g., frogs, elephants, man), reside inside a protective housing such as a cell membrane in the case of single-cell organisms or inside an outer skin (or shell) in the case of multi-cellular life forms (or an extra space suit in the case of astronauts). These outer protective coverings are designed to keep nasty things in the environment outside and life friendly things inside. Cell membranes have biologically controlled pores (or gates) that can be opened or closed to allow the organism to take in life critical chemicals or materials from the outside environment that they need to stay healthy and toss unfriendly chemicals (e.g. metabolic waste products) to the outside world. Higher multi-cellular life forms, such as man, use an additional method to take in needed materials or chemicals from the outside world and toss away unwanted stuff to the outside (i.e., eating/drinking and going to the potty).

In addition to the multitudes of basic life supporting bodily functions that nervous systems assist their owners with, all nervous systems also control what we humans somewhat loosely refer to as mental functions. In addition to the nervous system ordering the release of special chemicals into the blood stream that triggers specific glandular or muscular activities that control such bodily functions as eating, reproducing, flight from danger (running), etc., nervous systems also allow their owners to perform a multitude of more complex sensory functions such as seeing, hearing, smelling, feeling (e.g., skin touch) as well as higher order mental functions like remembering (memory), perception (recognition), problem solving, and that other kind of "feeling", (i.e., emotions), etc. As we ascend the evolutionary ladder from single-cell life (e.g., bacteria, amoebae) to more and more complex multi-cellular life (e.g., amphibians, reptiles, mammals, and man) the relative proportion of these higher level mental functions apparently increase markedly. I say "apparently" because biological and life scientists have been somewhat slow in recognizing the existence of certain higher level mental functions in some animal species. For example, we now know that problem solving, tool making, and certain kinds of symbolic communication (pre-language) abilities are also present in many of our mammalian (e.g. dolphins, primates) and even non-mammalian (e.g., some birds) relatives. Certain primate species (e.g., chimpanzees and gorillas) are known to be quite good at using simple tools to solve problems, as well as symbols (sign language) for communication. We also know that dolphins, porpoises, and even whales talk to each other constantly (using strange whistling or clicking sounds), even though our scientists do not have a clue as to how to break their language code. Thus, higher level mental functions (e.g., the ability to problem solve and communicate with other members of their own species) is not unique to humans. The fact that the possession of higher level mental functions is not unique to man reinforces the idea that the evolution of humans was not necessarily a rare fluke of nature (or even the manifestation of a unique higher creative power) but is probably a common outcome of the normal evolutionary process, if the process is given enough time (and environmental resources) to happen.

All Nervous Systems on Our Planet, in Spite of Looking Different from Each Other, Appear to Reflect a Single Origin or Be "Cut from the Same Mold"

The basic building block from which all biological nervous systems on Earth are constructed is a microscopic entity known as the *neuron* or *brain cell*. The simplest or most primitive nervous systems contain hundreds or thousands of these basic structural units, while more advanced species may contain many millions or even billions. Table 3.1 lists the total numbers of brain cells or neurons that a representative sample of animals ranging from primitive to advanced species are believed to contain. Since all animal life on Earth evolved from the same common ancestor, all organisms from the smallest multi-cellular organisms (e.g., earthworms, snails, lobsters, insects, etc.) to octopi (or octopuses), elephants, and man have nervous systems that are remarkably similar in terms of their basic construction and function. As the old saying goes. . . "*a nerve cell is a nerve cell, which is a nerve cell*"! The structure of individual brain cells or neurons is basically the same all across the animal evolutionary spectrum. Figure 3.3a, b shows drawings of what typical nerve cells from the brain of almost any animal species would look like. All nerve cells, whether belonging to snails, frogs, or man, or almost any other animal species, are constructed in the same way and work in the same basic manner. All neurons or brain cells have fibers called **dendrites** that conduct electrical impulses towards the **cell nucleus** and then away from the cell nucleus along a second kind of fiber called an **axon**. When the electrical impulse reaches the end of the axon, one of several events can happen next. If the impulse reaches a gland, muscle, or some other biological structure in the body, it may cause the gland to secrete some kind of chemical (e.g., a hormone), or the muscle to contract or relax, or trigger some other kind of bodily (or mental) function. If the nerve impulse needs to go a greater distance inside the brain or body to do its job, it may need to transfer itself to other nerve cells in order to continue its journey. When the nerve impulse reaches the end

Table 3.1 Total numbers of neurons that are believed to exist in brains of members of representative animal species ranging from roundworms to humans

Worms	Leech	Snail	Fruit fly	Cockroach	Frog	Mouse	Octopus	Man	Elephant
300	10,000	11,000	100,000	1 million	16 million	75 million	300 million	100 billion	200 billion

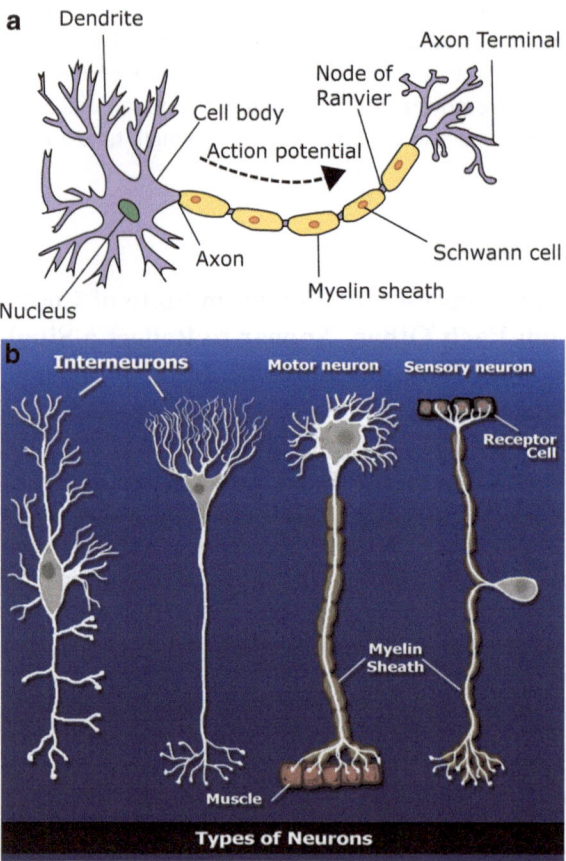

Fig. 3.3 (**a**) Shows an artist's drawing of a typical nerve cell from a more advanced species of animal. The numbers of individual branching dendrites and axon terminals range widely among different nerve cells in different parts of the brain and also in different animal species. The nerve cell shown here has fatty insulating substances called myelin sheaths wrapped around its axon. These myelin sheaths tend to occur in more advanced species and in parts of the brain that require faster nerve transmission or speed (image credit: U.S. National Cancer Institute's Surveillance, Epidemiology and End Results Program). (**b**) Shows that while all neurons have the same basic parts (axons, dendrites, cell nucleus, etc.) the parts may be put together differently depending on whether a particular neuron is an interneuron (i.e., transfers impulses from one neuron to the next neuron in a chain of neurons) or is a sensory or a motor neuron (image credit: Gerry Keegan)

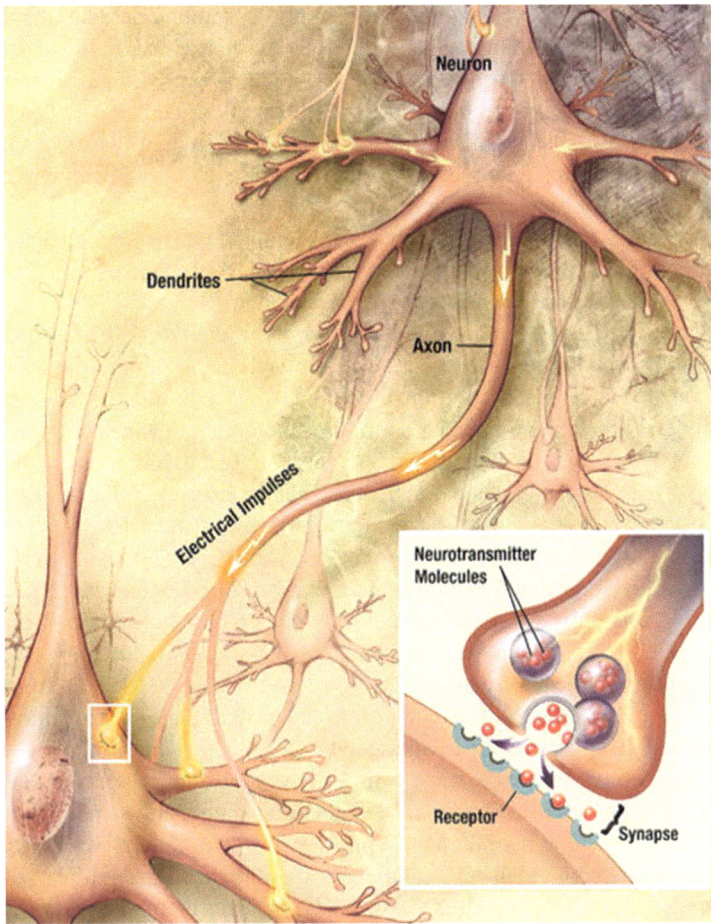

Fig. 3.4 Shows a more "life like" drawing of how neurons and synapses work in animal brains (image credit: U.S. National Library of Medicine/www.nlm.nih.gov)

of one axon it will encounter a special structure called a **synapse**. A synapse (Fig. 3.4) is a structure where the nerve impulse in the current nerve cell ends (or is stopped or "deactivated") and a new nerve impulse is generated and starts up in the next nerve cell in the chain. The impulse from the first neuron causes a special chemical, called a **neural transmitter**, to be released into the space (called the **synaptic cleft**) between the end of the first axon and the dendrite of the next nerve cell. This chemical then causes an electrical impulse to be triggered in the next nerve cell's dendrite which then begins traveling towards this new cell's nucleus and then to its axon. Although axons are long thin fiber-like objects, they are almost never long enough to allow the brain to transmit nerve impulses all the way to distant organs or muscles in the body or for these structures to directly send impulses back to the brain (hence the author's "highway or freeway" analogy).

All nerve impulses, in even the smallest or most primitive animals, have to be transmitted (or relayed) from one neuron to the next neuron via this dendrite to cell nucleus, to axon, to synapse, to dendrite.... mode of nerve conduction. So, one neuron "talks" (sends an electrical impulse) to the next neuron in a long chain, which then talks to the next neuron, until the nerve impulse finally reaches its destination, i.e. a gland, an organ, or another region of the brain, where some kind of bodily activity or mental function is triggered.

The above description is the author's best attempt to convey to the reader, in the simplest possible manner, how the nervous systems of all animals on Earth work. However, it is far too simplified to even begin to do justice to the extreme complexity with which even the simplest nervous systems work. Just as describing how a row boat works fails to accurately convey how an ocean liner works, my description also falls far short of its goal. I will now need to "muddy the waters" and risk completely boggling the reader's mind by attempting to expand on this simple description. While, in all animal species, nerve cells are composed of the same basic parts, i.e. dendrites, axons, cell nuclei, and synapses , and information is passed from one nerve cell to the next in the same manner, i.e., by triggering electrical impulses that are passed from one neuron to the next, the typical animal nervous system (see Table 3.1) may contain many millions or billions of individual nerve cells. Although each nerve cell has only one cell nucleus (which controls the critical metabolic life functions of the nerve cell, plus, most importantly, directs the manufacturing of special chemical transmitters that are needed to trigger and transmit nerve impulses from one neuron to the next), **many neurons have more than just a single dendrite and a single axon**. While the human brain is estimated to contain 100 billion neurons, most neurons have many short branching dendrites attached to the neuron's cell body and a single long axon that, when it gets close to the next neurons in the neural chain, branches into a number of **axon terminals** (Fig. 3.3a). In many neurons, the numbers of these branching dendrites and axons may number in the tens, hundreds, thousands, or even more. The fact that most neurons in the brain have multiple branching dendrites and axons means that they are able to receive inputs (electrical impulses) from many other neurons and also send nerve impulses to a multitude of other neurons. Thus, it does not seem to be an exaggeration to state that neurons in a typical animal brain are *well connected* (or inter-connected) to each other. So, how many axon and dendrite fibers does the average animal brain actually contain? No one knows for sure, but, for more advanced species like us, the number is probably well into the trillions! Some parts of the brain have neurons with incredible numbers of branching axons or dendrites (or both), while other parts of the brain may have neurons that host far fewer branches.

Now, for another mind boggling piece of information! The author said earlier that the manner in which information is passed from one place in the brain to another is via electrical impulses that are transmitted from neuron to neuron along dendrites and axons. While the chemicals that trigger nerve impulses (i.e., the nerve transmitters) may be different in different parts of the brain (i.e., chemically different and even called by different names, e.g., Dopamine, Adrenaline,

Acetylcholine, Epinephrine, etc.), there are only two major functional types of neural transmitters in animal brains. One type is called an **excitatory** transmitter, and the other type is an **inhibitory** transmitter. With some nerve cells, when the electrical impulse reaches the end of an axon it will cause an excitatory chemical transmitter to be released into the synaptic cleft which then causes a new nerve impulse to be triggered in the dendrite of the next neuron. *This is the mode of neural conduction that I was totally focused on up to this point in the book.* Nevertheless, with other neurons in all animal brains you frequently get the exact opposite effect. With this second kind of neuron, another kind of neural transmitter is released into the synaptic cleft which acts to inhibit the triggering of an electrical impulse in the next neuron. Thus, some kinds of neurons cause nerve impulses to be transmitted on to the next neuron (or neurons) in the chain of nerves, while other neurons actually inhibit or prevent nerve impulses from being triggered in the next neurons in the chain. *Thus, in a very real sense, the brains of animals on Earth mimic our computers. Information transfer in computers is digital, involving either an "on" or "off" electrical state. Our brains seem to work much the same except we call it excitatory or inhibitory neural information transfer instead of "digital computer processing".* Although this amazing similarity in the structure and function of our brains and our computers is probably incidental, the fact remains that many of mankind's most significant discoveries or inventions can be directly linked to some kind of preceding "invention" by mother nature. The similarity between airplane wings and bird's wings is probably no accident as are many of our other remarkable inventions which man borrowed or copied (oftentimes without giving full credit) from nature's own solutions to life's environmental challenges.

Unfortunately, the author will now have to again complicate the reader's understanding by again raising the bar on how complex animal brain function can be. When electrical activity from several surrounding nerves merge onto the dendrites of another neuron, whether or not a new nerve impulse is triggered in this next nerve cell may be dependent on the numbers and actual timing of the arrival of excitatory and inhibitory neural inputs to this neuron. If most of the inputs to the first neuron are excitatory, then a nerve impulse which, by the way, are frequently referred to by brain scientists in their textbooks as *action potentials* will be triggered in the next neuron.[3] Likewise, if most of the inputs to the first neuron are inhibitory, no nerve impulse or action potential may be triggered in the next

---

[3] In order to avoid boggling the reader's brain any more than is necessary, the author has deliberately chosen not to attempt to describe the complex process by which nerve impulses or *action potentials* are actually created and transmitted. The process involves the transfer of positively and negatively charged potassium and sodium ions across the coverings (outer layers or "skin") of the cell membranes of nerve cells, axons, and dendrites and the chemically induced "all-or-none" triggering (or inhibition) of new nerve impulses in other neurons. If the reader desires to learn more about this quite complex process, I would suggest consulting almost any recent introductory level textbook on neuroanatomy or neurophysiology. I have referenced three such introductory level books (Amthor 2012; Dubin 2002; Liebman 1986) in my Further Suggested Readings section at the end of the present book.

nerve cell. However, if a mix of both excitatory and inhibitory inputs arrive at the first neuron, whether or not a new action potential is triggered in the next neuron may be dependent not only on the relative numbers of the two types of inputs (i.e., whether more inputs are excitatory than inhibitory, or vice versa) but also on the relative temporal timing of the arrival of the two types of inputs. If some of the arriving inputs, whether excitatory or inhibitory, are either early or late in arriving, they may not be able to influence the triggering (or non-triggering) of an impulse in the next neuron. (*Could this be the physiological bases of our occasional "memory lapses" or even "brain storms"?*—food for thought!).

In the last paragraph, the author described how the presence of either excitatory or inhibitory chemical neural transmitters at the synaptic junction between two nerve cells either triggers a nerve impulse or prevents (inhibits) it from being triggered in the next neuron. This description of how nerve impulses are either triggered or inhibited in the brain is an over-simplification that I will need to once again clarify in order to keep from confusing the reader even further. The reason that the temporal timing of the arrival of excitatory or inhibitory neural inputs at one neuron is so important in determining whether or not another neural impulse will be triggered in the next neuron, is that nerve cells, as part of their inherent nature do not just transmit single nerve pulses (action potentials) every so often, but are typically continually firing nerve impulses one after the other, in a continual repeating fashion. It seems that, in normal brains, most nerve cells tend to repeatedly fire at a specific rate even when they are not actively involved in some kind of behavioral or mental activity. This ongoing firing of "resting" nerve cells when they are not "working" (which is analogous to the idling of a car's engine) is referred to by brain scientists as the particular nerve cell's spontaneous firing rate. *Remarkably, it seems that it is when the firing rates of nerve cells either increase or decrease in frequency (rather than starting or stopping) that information is being transferred from one part of the brain to another part.* Thus, most neurons are constantly building and deactivating neural transmitters virtually all the time. After a new nerve impulse is "triggered" (i.e. spontaneous firing rate is either increased or decreased), the nerve cell immediately deactivates (neutralizes) the existing transmitter and sends a message to the nerve cell nucleus to immediately build new transmitters and move them to the synapse so they can be available to trigger (or inhibit) the next nerve impulse when needed.[4] Nerve cells in all animal brains typically repeat this process (neutralizing old transmitters, building and activating new transmitters) over and over again very rapidly, perhaps tens, hundreds, or more times per second. This very busy lifestyle of neurons is the reason our brains are the most greedy parts of our bodies with respect to needing food to keep on working. This constant need of neurons to repeatedly go through the amazingly fast and

---

[4] In reality, rather than neurons "ordering" the cell nucleus to build new neural transmitters and ship them to the synapse to trigger a new nerve impulse, the nerve cell nucleus contains small organelles (factories) that are constantly manufacturing neural transmitters and placing them into small packets to be delivered down the axon to the synapse so that they will be readily available whenever a new action potential needs to be triggered (or inhibited).

complex process of creating new nerve impulses also makes us vulnerable to any disease process that interrupts the normal rhythm of this process. Medical problems such as Parkinson's disease and other forms of neuromotor disorders interrupt this process and cause uncontrollable muscular twitching and coordination problems, plus possibly higher level mental functions such as memory loss or sensory hallucinations.

As a direct result of 40 some years of intense reading, studying, and experimenting/publishing, the author's brain is unfortunately crammed full of much more information about how brains work than that contained in the preceding paragraphs. Based on what I have already told the reader about how animal brains work, you may feel that the typical animal brain is, indeed, extremely complex. Nevertheless, I am here to tell you that my description of our Earth-bound nervous systems has only skimmed the bare essentials of this complex topic, and many more books would be needed to finish the job. (***Later in this book, the author will describe how our computer science colleagues are now beginning to assist us neuroscientists in solving this incredible and very challenging task***.) For example, while man's brain is estimated to contain at least 100 billion individual nerve cells, this is only the tip of the iceberg, so-to-speak. For every brain cell or nerve cell, some animal brains may also contain as many as ten other kinds of cells that the neuroanatomists call **glial** cells.[5] Neuroscientists do not know everything that glial cells do to assist nerve cells, but they undoubtedly have a multitude of supporting roles including physical (holding the nerve cells in place), supplying nutrients, insulating neurons electrically, destroying pathogens that could harm nerve cells, removing dead neurons, and providing guidance in directing the growth of axons to their dendritic targets (important for brain development and possibly repair following injuries). In a very real sense, the glial cells probably support the nerve cells in a manner not unlike what the members of "pit crews" do in supporting race car drivers in racing competitions.

---

[5] More recent studies have started to dispute this 10 to 1 glial cell to brain cell ratio. Some studies now suggest the ratio might be closer to 1 to 1. The ratio could, however, vary widely among different animal species, or between different parts of the brain. Counting numbers of glial cells in different brains does not appear to be one of neuroscience's burning questions! We neuroscience professors have never been able to interest a doctoral student into taking on this project for his/her doctoral dissertation (thesis).

Could Extraterrestrial Nervous Systems Be Radically Different from Each Other as Well as Function in Totally Different Ways?

The discovery of extremophiles living all over our planet plus the recent findings by the planet hunters that our sun is far from being alone in its ability to host a multitude of orbiting planets that could be potential homes for alien life is now triggering a growing number of scientists to believe that life in the universe may be both common plus quite diverse in terms of its biological makeup. Most scientists now believe that there are two critical events that absolutely must occur in order to make it possible for biological life to successfully evolve and survive in planetary environments anywhere in our universe. First, all creatures must, via some process equivalent to what we call biological evolution on Earth, develop external morphologies or body characteristics/structures and internal physiological processes that allow them to adjust to and live in a variety of different external environments. Such life forms must, to site a few possible examples, develop long enough fur (or other external body coverings) and/or internal heat generating metabolisms to allow them to stay alive no matter how cold or hot their world may be. If they live on a totally water covered planet (water world), they must develop fins or other means of locomotion that are suitable for their environment. If they live on a large planet which has a strong gravitational field, they must develop sufficiently strong bodily structures that allow them to walk, crawl, or engage in other kinds of surface locomotion, or large enough wings to be able to easily fly in what may be a thicker (or even stormier or turbulent) atmosphere. If they live on an extremely cold ice world, they must develop means (e.g., anti-freeze like chemicals) to allow them to be mobile and not freeze. To make a long story short, all living creatures, no matter what kind of strange, hostile, or unfriendly kind of world (as defined by us earthlings, of course) they may live on, must be capable of undergoing evolutionary changes to successfully cope with whatever their own local, *and probably itself evolving*, environment may throw at them.

And, secondly, all living creatures must develop some kind of central biological control system that can facilitate and manage all of the internal bodily processes (e.g., digestion, growth, muscle movement, reproduction, etc.) that are needed to keep them alive and well in their current environment, as well as successfully respond to any threats to their survival that may result from changes in their immediate external (or internal) environment. And, although not directly linked to their day-to-day survival, living creatures, in order to live longer and better must be intelligent (once again, as defined by our own species) in being able to also develop biological mechanisms that allow them to learn and store information somewhere in their bodies about what is presently happening to them in their current environment so they, whenever necessary, will be able to use this information to respond appropriately to facilitate their lives or to avoid injury or destruction.

And, of course, on Earth, all of these requirements for life are controlled by something our scientists call a "brain" or a "nervous system". On our small rocky water planet everything the life form we call an animal does, sees, hears, thinks, learns, remembers, figures out, etc., in terms of its daily life-related activities, is controlled by billions or trillions of biochemically generated electrical impulses that are constantly racing along the dendrite and axon fibers of the millions or billions of nerve cells that are located inside their pea-size to cabbage-size brains.[6] The job of our brain scientists is to figure out how this huge and incredibly complex melee of electrical and chemical "goings on" in our brains gets translated into the sights, sounds, perceptions, ideas, problem solutions, memories, temper tantrums, ad infinitum, that our brains come up with each and every day of our lives. While scientists think they know a considerable amount about how the simplest activities of our brains may work, such as how a knee jerk is triggered by hitting the knee cap with a small rubber mallet, they are, at least at the present time, totally clueless as to how an Albert Einstein develops a theory of relativity or an infant learns to recognize its mother's face.

Therefore, a very important function of any nervous system, in addition to controlling its owners normal day-to-day biological functions (eating, sleeping, reproducing, physical growth/development, etc.) is to allow the organism to *mentally internalize* (within its brain) some kind of chemical or biological model or representation (electrical, chemical, physiological memories) of the external world (plus also its internal world, i.e., what is going on within its own body) so that it can allow the owner to quickly and accurately adjust to any physical or chemical changes inside their own body or in the world around them that is needed to ensure their survival and ability to reproduce. This internalized modeling process must be semi-permanent so that organisms can form, and place into storage somewhere in the brain, memories (images) of past environmental changes that can be readily available to assist them in responding to new changes. At least on our planet, animals must be capable of sensing (being "aware" of) events or changes in their present environment, such as seeing (sensing the light reflected off other objects), hearing (detecting vibrations in air or water molecules triggered by chirping birds, twigs snapping, or predatory fish, etc.), smelling or tasting (sensing odors with their noses, or chemicals with their tongues), and feeling (i.e., skin touch or pressure, including pain).

In order to survive and successfully adjust to future environmental changes, animals must learn (and remember) an amazing amount of information related to what psychologists refer to as the "cause and effect" relationships between almost everything that happens inside their own bodies and in the physical world around them. Primitive organisms (and also advanced organisms) must learn (i.e., sense or

---

[6] How fast do nerve impulses travel, you ask. The speed at which action potentials travel along typical axons or dendrites varies greatly, depending on the specific kind of nerve involved (i.e., whether the axons and dendrites have fatty myelin sheaths wrapped around them which speeds up neural conduction or do not have myelin sheaths which slows down their conduction), or the particular species involved, i.e., primitive or advanced (e.g., cockroach or man, etc.). The typical range of speed varies from a typical "snail's pace" of 2 miles/h (in snails or other kinds of insects, of course) to 250 miles/h in some animal brains (cats, birds of prey, jack rabbits, etc).

perceive the relationships among the different events and store this information in their bodies for later access) that certain potential food items that look, smell, and taste in a certain way may produce (cause) the effect of digestive distress if eaten, or that other creatures that look a certain way (e.g., bears) may attack them and cause them to be eaten (the *effect*), More advanced organisms must learn that an automobile that is heading toward them may "cause" the "effect" of their being hit and killed. And even more advanced organisms, like man, must learn tons of cause and effect relationships in the world around them. Rocket scientists must learn that making a satellite that is circling the Earth move too slowly may cause it to be pulled back to Earth by gravity and crashing.

Intelligent extraterrestrials on other planets must also be capable of filling their nervous systems (or equivalent structures) with incredible amounts of stored information related to the cause and effect relationships between themselves and objects or physical events on their own world. Hearing and seeing may or may not be universal requirements for survival on other worlds and other environments might require their inhabitants to develop systems that are sensitive to other kinds (wavelengths) of light or sound vibrations, or even totally different kinds of physical stimuli that we humans may not consider important or, as yet, even know about. Everywhere on our own planet, some animal species have developed sensory systems that allow them to respond to sounds and sights that we humans are totally insensitive to. Some of our own migrating animal species (birds, butterflies, turtles, sharks, and whales) may even use subtle changes in the Earth's magnetic field to assist them in navigating or migrating. Bats utilize echoes from extremely high frequency sounds that they themselves generate to allow them to avoid flying into objects in dark caves, and elephants are now known to use extremely low frequency sounds to communicate with other members of their own species over very long distances. The evolutionary process on Earth has allowed a multitude of other animal species to develop sensory systems that take advantage of sound, light, or other environmental stimuli that are readily available to them in their home environments but not in ours. Many deep sea creatures have even been discovered to communicate with other members of their own species by producing complex and colorful displays of light stimuli that their bodies are able to chemically create (i.e., "bioluminescence" or the ability of some organisms to generate light stimuli). Certain species of fish that originally had eyes and vision, slowly lost their eyes when they began living in the waters of dark caves where there was no light available. Since they no longer needed eyes to see, evolution allowed them to get rid of their eyes to conserve their metabolic energy requirements.

Thus, ETs on other planets with totally different environments may develop sensory systems that are completely different from what we humans use here on Earth. What we humans loosely describe as "extrasensory perception" (ESP) or even "mental telepathy" may be very common on many other worlds. Also, the transmission of information in some alien nervous systems may not involve electricity and the movement of positively and negatively charged ions (potassium and sodium atoms) across nerve cell membranes but perhaps the transmission of radio frequency or other forms of electromagnetic, chemical, or even physical energy.

Life on our own planet Earth may be even more unique than any of us could ever imagine. ET may not only look totally different from us but may also act, think, and interact with each other and their surrounding environments using different physiological (or mechanical) systems that utilize totally different kinds of stimuli. The idea that our future astronauts will encounter ETs that they can look in the eye, shake hands or appendages with and even talk to is probably unlikely unless panspermia turns out to be more common in the universe than we now believe is possible, or the laws of evolution (*especially those related to our Earth-bound forms of "convergent" evolution*) turn out to be more universal than we now believe is likely or possible. It is possible that the idea of many of our scientists that intelligent life is extremely rare in the universe may be more due to the fact that intelligence, instead of being rare or uncommon, may be quite common but exist in different or unique formats on many other worlds that we humans would not be able to recognize, or possibly be unable to interact with. And even the most bizarre and "far out" members of our science fiction writing associates would likely fall into this trap. While intelligent alien grasshopper-like creatures might sell books and movie tickets, intelligent interstellar gas clouds or oozing ("cruising") tar pits might not!

The very real possibility that alien nervous systems on other worlds could be totally unique and entirely different from the *incredibly consistent* format that such systems assumed on our own planet may go a long way in accounting for the Fermi Paradox (Webb 2002).[7] ETs on other worlds may not only think and act differently from us, they may also sense their surroundings using totally different physical (or chemical) stimuli than we do. SETI (The Search for Extraterrestrial Intelligence)'s "failure" thus far to make contact with ET may simply be that many ETs may have nothing in common with us to talk about. And, while the more intelligent ETs may and most likely do share our knowledge of the physical world around them, i.e. the so-called universal laws of physics and chemistry, they may have, in addition to different sensory means of detecting or sensing such things, totally unique means or techniques of sharing this information among themselves which would make no sense to us carbon-based life forms. Mankind's concept (or definition) of intelligence may, in the universe as a whole, be "parochial" (restricted or rare). Thus, in a very real sense, mankind may be virtually alone in the universe! Instead of talking to ET, we might need to learn how to flash colorful light displays at him, her, or it, or whatever.

---

[7] In 1950, the Nobel prize winning atomic physicist Enrico Fermi was having lunch with a small group of his science colleagues. The group was having an informal discussion related to the possibility that large numbers of habitable planetary systems might exist in the universe, and some could be home to advanced technological civilizations. At the end of this discussion, Dr. Fermi reportedly turned to the person seated next to him at the table and said something to the effect of "*So, where is everyone?*" While a very simple question at the time, Dr. Fermi's question has, in subsequent years, become the famous **Fermi Paradox** that is at the center of the SETI research effort. If there are so many potentially habitable worlds out there, why has science so far not found some evidence of the existence of intelligent extraterrestrial civilizations?

And one more, and even more bizarre possibility, is that intelligence on other worlds may not be the sole province of single individual members of a living species. On Earth, although the reader's brain and my brain developed in a virtually identical or similar fashion during gestation and following birth, our brains are almost entirely independent of each other in terms of many of their functions. While we can inter-communicate (talk) and exchange thoughts and ideas with each other, this communication interchange is notoriously slow and inefficient, and frequently subject to error. This problem is very obvious whenever groups of people gather to "put their heads together" to make common plans or solve common problems, or just socially chit-chat. If you do not believe this, just tune into a television broadcast of the U.S. Congress in one of its law making sessions. In my long career as a university professor, my most unproductive times were when I attended faculty meetings. It is possible that alien species on other worlds may have developed nervous systems that function not as independent isolated entities but as individual members (physically separate units or components) of whole groups of functionally inter-linked or connected brains, each of which has a specialty in terms of its individual function(s) which can be combined with the functions of other specialized brains to produce a combined group form of "intelligence". Many members of the most dominant form of animal life on our planet, i.e., our own insect species, exhibit this unique form of group intelligence. Ants and bees, for example, have queens, workers, soldiers, and other specialized members which, when put together, form more complex (intelligent?) ant colonies (or bee hives). This kind of group intelligence (or "swarm" intelligence) may actually be the dominant format that intelligent life on some other worlds may have assumed. Some serious scientists are today beginning to speculate that the onset of the world-wide web or internet may be the opening salvo of mankind's eventually transitioning to some kind of "artificial" group intelligence. This form of intelligence would, of course, be extremely beneficial to us humans, although quite bizarre and virtually impossible to imagine.

However, as for the next stage of human evolution, i.e., the transition from biological brains to "intelligent machines", mankind may already be well on the road to achieving this type of group intelligence in the not so distant future. Individual computers all over the world are now rapidly being linked together to produce incredibly more efficient and speedy super-computer or parallel computer systems that can achieve far more than individual computers. Mechanized group intelligence (or "artificial" intelligence) now seems to have been launched on our planet—how far this transition will go and how beneficial it will be to our human lives is still to be determined, but will likely be determined (for the better or worse) in just a few short years. This transition from individual biological intelligence to forms of artificial group intelligence may have already occurred elsewhere in our vast universe. It is also possible that, in the far distant future, as computer systems get smaller and smaller and more sophisticated that such artificial intelligent devices could be implanted inside (or "plugged" into) our biological brains or those of ETs to allow their owners to turn them on whenever they want to engage in some kinds of intelligent group functions with their neighbors or far distant colleagues.

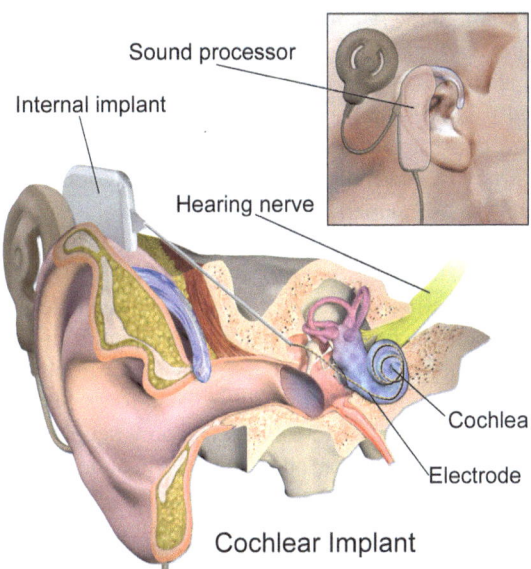

Fig. 3.5 *Shows an artist's drawing of how some scientists <u>can now "plug" a computer into a</u> <u>human brain to control (or assist) brain function</u>*. This statement is not meant to be science fiction! Clinical audiologists have, for the past 15 years, started doing just that. Since the author is a clinical audiologist, I am very familiar with this clinical procedure. All of us, if healthy, are born with an auditory nervous system which allows physical sound energies (produced by vibrating air molecules) from the outside world to travel through the eardrum into the cochlea where specialized cells (called hair cells) are stimulated that send special biochemical signals to the brain which allow patients to hear sounds. In some patients, the sensory hair cells in the cochlea are damaged and do not function properly. However, these cells can be made to work if a computer-based device called a *cochlear implant* along with a special *electronic sound processor* are surgically placed inside the skull just behind the patient's pinna. The sound processor detects the vibration of air molecules that occurs whenever sound is presented. The cochlear implant then converts this information into special electrical signals that are sent to the cochlea to artificially stimulate the hair cells and allow the patient to hear (image credit: Wikipedia Commons)

Although mankind cannot yet insert small computers into the brains of humans to improve their functional capabilities, today's computer technology has reached the point that scientists can use such implanted devices to restore many types of sensory and motor function in damaged human brains. The present author, as a clinical audiologist, is very familiar with one clinical technique that is able to restore normal auditory function (hearing) in some human patients (Fig. 3.5). This is obviously the first step in man's eventually being able to implant tiny computers into human brains to turn them into "super brains".

The author, therefore, having devoted 40 years of his life to trying to figure out how the nervous systems of living creatures on Earth works, must admit that he now finds himself virtually clueless as to how he believes the nervous systems of extraterrestrials on other worlds might evolve and how they might work. As a working brain scientist, I find the nervous systems of even the most primitive animals on our planet to be very complex; and the brains of more advanced species are beyond complex and some actually border on the bizarre. I personally find it absolutely amazing that, in spite of this extreme range in the complexity of the brains of so-called primitive and advanced animals, the brains of all living specimens on Earth are constructed and work in a remarkably similar fashion. This similarity in the structure and function of the nervous systems of different animal species on our planet stands in marked contrast to the absolutely incredible diversity in the external and other internal physical morphologies of different species.

And even more amazing, especially for our astrobiologists, is how a biological process called evolution allowed such huge numbers of complex intelligent organisms on our own home planet to survive and prosper in an incredibly hostile environment that was subjected to at least five major catastrophic extinction events that could have destroyed all life at any time. Therefore, this author, along with rapidly growing numbers of his fellow scientists, now believe life on Earth is not only extremely complex and diverse, but is very resilient and flexible in being able to pop up and survive in the harshest of environmental conditions. Now that our scientists have started finding exoplanets and other solar systems everywhere in our own galaxy, how rare can life be out there? This author, for one, now believes that advanced forms of intelligent life, while relatively rare mathematically compared to simple life forms (e.g., single-cell microbes), are probably far more diverse in terms of their physiological and behavioral manifestations than most of us (scientists as well as non-scientists) can possibly imagine. In the next chapter, I will take off my professional brain hat and put on my amateur astrobiology hat and return to discussing those other equally important but complex topics related to this book of how life may evolve on other worlds in our universe and perhaps even survive long enough to transition into that complex thing that we Earth-bound life forms loosely refer to as "intelligent living creatures".

Chapter 4
Evolution of Intelligent Nervous Systems on Other Worlds in the Universe

As indicated in the last chapter, while many scientists (including the author) know a considerable amount about how nervous systems work on our own planet, none of us has any knowledge of whether or not intelligent life might be able to evolve on other worlds in the universe. Therefore, I will need to "stay the course" once again in the present chapter and begin by providing the reader with a simplified overview of the only example of how biological evolution works anywhere, and that is how our scientists believe it works on Earth (Baross 2007; Charlesworth and Charlesworth 2003; Krukonis and Barr 2008; Futuyma 2005; Scientific American Reader 2006; Schopf 2002). Then, in the second part of the chapter, I will return to the incredibly fun goal of the present book of trying to apply my knowledge of how nervous systems evolved on Earth to speculating about how comparable systems might develop elsewhere in the universe (Benner and Davies 2012; Dreamer 2011).

How Biological Evolution Works on Planet Earth

A little over 150 years ago, a British naturalist named Charles Darwin told us that, in order to survive on our planet, all living creatures had to adjust or adapt themselves to long term changes in the environment in which they lived. On November 24, 1859, Darwin published his landmark book entitled *The Origin of Species* (full title of "*On the Origin of Species by Means of Natural Selection on the Preservation of Favoured Races in the Struggle for Life*") which completely changed the course of the life sciences. Although his theory of evolution and natural selection has turned out to be amazingly accurate, Charles Darwin knew virtually nothing of genes, mutations, or any of the physiological or biochemical underpinnings that would almost a century later provide the scientific proof for his remarkable brainstorm. The science (which is no longer a theory) of evolution is based on two major observations related to the way in which life forms change over time. The first observation relates to the fact that most of the time, and in most

© Springer International Publishing Switzerland 2015
J.L. Cranford, *Astrobiological Neurosystems*, Astronomers' Universe,
DOI 10.1007/978-3-319-10419-5_4

environments, the breeding and reproduction of individual members of a specific animal population (defined as all of the individuals of the same inter-breeding species that live together in one geographic location) occurs at a faster rate than the local environmental resources can support. More offspring are produced than can be adequately fed or sheltered. This results in competition among the individual members of the population. Thus, the population members are forced into a *struggle for survival*. The second observation relates to the fact that, in all animal species, the individual members of the population exhibit differences among themselves with respect to specific morphological and functional characteristics. Some are larger, stronger, can run faster, can see or hear better, have longer fur (or shorter fur), and so on. Almost all of these characteristics are inherited from their parents. A very small number of these differences (i.e., the **mutations**), however, are not. Whether inherited or not, some of these characteristics will give some individual members of the group an advantage over the others in terms of being able to survive long enough to breed and pass their "lucky genes" on to their own offspring. Other members of the group may have inherited characteristics that place them at a disadvantage in this survival of the fittest saga. Therefore, in the 1850s, when Darwin was tackling the daunting issue of the evolution of life, he could only address the question of "what" happens; the question of the "how" would take another 100 years or more to begin to be resolved.

In 1953, the question of the "how" of this greatest of all biological mystery stories began to be solved when James Watson and Francis Crick finally broke the genetic code (Watson 1998)! At its very essence, the primary goal of all science is to *explain* things that happen in the natural world. In science, an explanation is a description of all the different *causes* that come together to produce a specific *effect*. Yet, the first stage of any new science is to totally and completely describe the details of *what happens* with respect to some phenomenon that we do not yet understand. This is what Darwin at first did. He observed many different animal species and described everything he could about them with respect to how they lived, where they lived, what they ate, and how their bodies were built. Darwin also studied the effects that controlled breeding (by animal breeders and farmers) had on rapidly accelerating the apparent rate of evolutionary changes in domesticated animals. Darwin's descriptive studies provided the critical "what happens" information that was needed for the new science of *Biological Evolution*. The second stage of any new science is to identify and isolate (separate) the *causes* and *effects* with respect to the phenomena being studied. The scientist's objective is now to determine *"what causes what"* and *"what are the effects of different causes"*. The detailed explanation or elucidation of the causes and effects related to biological evolution is where the twentieth century scientists, including Linus Pauling, James Watson, Francis Crick, and Lynn Margulis, plus many other distinguished scientists enter the picture. Before beginning a discussion of how intelligent life might evolve on other worlds in the universe that harbor different environmental conditions, I will provide a brief simplified overview of the genetic mechanisms that our life scientists believe underlies the phenomenon of biological evolution on our own planet.

As described above, we now know that the basic unit (i.e., the smallest indivisible entity) for the inheritance of biological traits (i.e., each of our personal physical or body characteristics, e.g., hair color, body size, racial background, sex, height, weight, etc.), are the individual **genes** that are located inside the cell nuclei of our body cells. Man has between 20,000 and 25,000 separate genes. Every cell of our body contains a complete set of these individual genes packaged in the 23 pairs of chromosomes housed in the nucleus of each cell (Fig. 4.1a, b). Now, how do "inheritable traits" change? The general term that scientists use to describe the occurrence of any new gene that has the potential to be passed along to its owner's offspring is a *gene mutation*. A gene mutation occurs whenever the construction or makeup of one of our genes is altered in some manner which results in a change in how it controls a specific behavioral or biological trait. For example, a gene might be modified in such a way that instead of allowing its owner to have long hair, it might cause baldness or, instead of being tall, to be short. There are many different types of causes for individual gene mutations. In the course of the replication of genetic material during growth or normal cell division, copying mistakes can occur. Still, for purposes of passing along inheritable traits, the only copying errors we need to be concerned with are the ones that take place in the organism's *gamete cells* (i.e., in the case of man and other animals, the sperm cells in the testes, or the egg cells in the ovary). Copying errors in other body cells may produce other undesirable effects such as unsightly moles, or skin cancer, etc., but these changes would not be passed along to the offspring. *It is important to note that, when we talk about genetic mutations, the vast majority of everyday mutations, whether they occur in body cells or gamete cells, are totally benign, i.e., have no observable effects on the individual organism's lifestyle, or mortality, etc.* When genetic mutations occur in gamete cells, while typically benign, they do have the potential to be passed along to and alter the future offspring's lifestyle, mortality, etc. In addition to simple copying errors, mutations can have a variety of other causes. Exposure to toxic chemicals from the environment, or in the diet, can produce mutations. Also, exposure to too much sunlight or radiation (X-rays, gamma rays associated with nuclear materials, etc.) can produce mutations. Once again, many organisms, including man, have evolved specialized enzymes that can frequently "make repairs" to such damaged genes. In recent years it has been discovered that certain types of mutations may even be *caused* by other genes that some unfortunate individuals may inherit from their parents.

When genetic mutations that occur in the gametes are passed on to the offspring, whether they have any effect on the individual organism's lifestyle will be determined by the specific environment in which the organism lives. If the altered gene is a "good gene" it may enhance its owner's health and other inherent biological functions in such a way that it will live long enough to breed and pass this new gene to the next generation. If it is a "bad gene" it may make the individual organism die prematurely or not be healthy enough to later court and breed. More likely than not, the altered gene is neither good nor bad and has no effect on the organism's current lifestyle, but might do so in the future. The altered gene could, therefore, produce biological changes that could either immediately facilitate or impede survival in

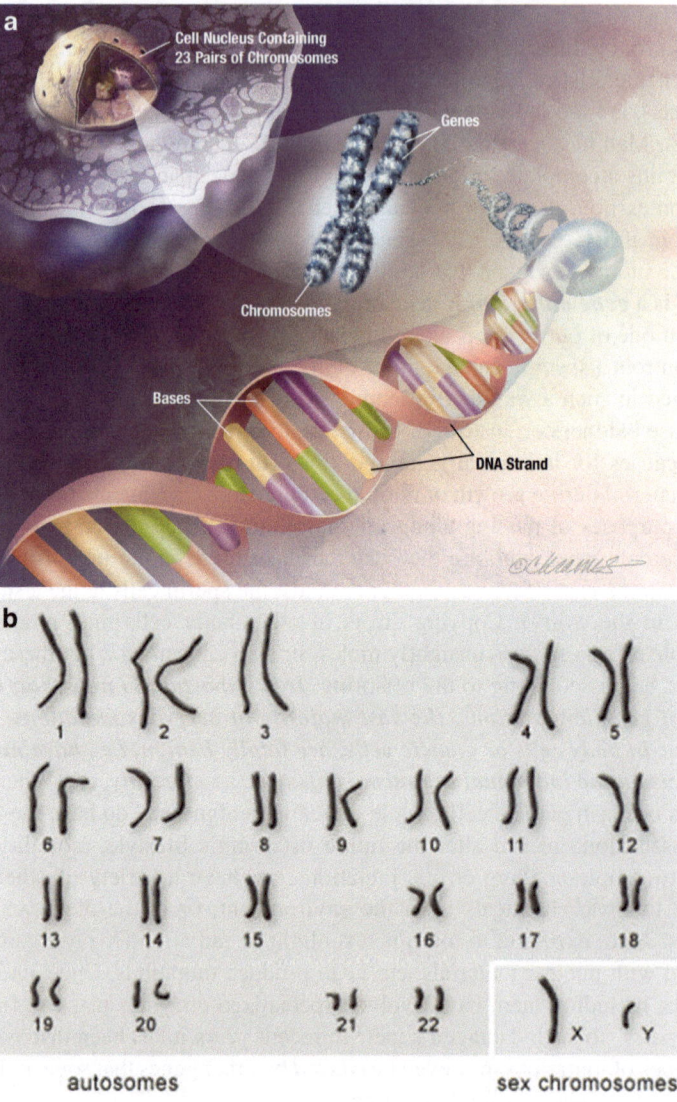

Fig. 4.1 (**a**) Shows a medical artist's drawing of how the complex "chemical instructions for building life" or DNA genetic materials are stored in each of the individual body cells of every animal species on Earth (image credit: NLM/NIH.gov). (**b**) Illustrates the differences in size and shape of each of the 23 pairs of chromosomes of humans (image credit: U.S. National Library of Medicine)

their current external environment, or, alternately, be available to do so when, in the future, the environment does change. *Therefore, the bottom line of Darwin's concept of evolution and survival of the fittest is simply the question of whether an inherited mutated gene facilitates an individual organism's ability to produce more viable offspring than their competitors who do not have the new altered gene*. Over time, as the numbers of group members with the new altered gene begin to outnumber those without the new gene, the new gene becomes a "normal" gene for the gene pool of the population.

The external environment in which a specific population of organisms lives is also frequently a major determinant of whether the occurrence of any given genetic mutation will result in evolutionary change. Slower environmental alterations are usually better in terms of permitting evolutionary change than are sudden alterations. Any new genetic mutation that, for example, might allow a given individual to adjust to a changing environmental condition (e.g., climate getting warmer, colder, wetter, dryer, etc.) will require some time (perhaps tens to hundreds of generations, or longer) to be able to spread throughout the population. If the climate change occurs suddenly (e.g. as a result of an asteroid impact or sudden intense volcano activity), the species may perish before it has time to incorporate the new protective gene into its genetic pool. Also, it appears that evolutionary change in a population can oftentimes be speeded up or enhanced when a small segment of a larger population becomes isolated. The smaller size of the isolated group allows the mutated gene to be dispersed to the majority of the group's membership in fewer generations. The rapid emergence of new species of organisms appears to be facilitated by any form of isolation event, be it climatic or geological. Perhaps the most dramatic example on our own planet of this sudden explosion of unique animal and plant species can be found in Australia. A little less than 100 million years ago, Australia and New Guinea separated from the supercontinent named Gondwana and remained isolated from the rest of the world for close to 40 million years. This allowed the evolution of many unique species of both animals and plants that differed dramatically from those found in other parts of the world, including kangaroos, wallabies, and the koala bear.

Different Forms of Biological Evolution May Operate on Exoplanets with Different Environments

Although Charles Darwin, in his famous 1859 book, made the idea of natural selection and evolution appear to be a relatively straightforward (although incredibly slow) process, the subsequent discoveries of hundreds of dedicated scientists all over the world has completely changed this simplistic view. It now seems that the scientific field of biological evolution may be just as mysterious and complicated as are the fields that underlie both the nervous system and the universe. Evolution is much more than genetic variations/mutations, survival of the fittest, and adaptation to changing environments. Darwin's principle of "natural selection", while very powerful and important to how life changes over time, is definitely not the only cause of the manner in which living organisms change or adapt to either short or long term modifications in the physical environments in which they live. It is very likely that Darwin's principle of natural selection is occasionally overshadowed by a multitude of other biological, chemical, and physical factors, including those that originate both within (e.g., geological, climatic) and beyond our own or any other planet (e.g., radiation from exploding stars, asteroid strikes) that collectively determine how, when, and where life changes may or may not occur, including even how quickly or slowly they occur. In the past 150 years, our geologists and paleontologists have uncovered a large amount of new evidence that the early Earth was plagued by a multitude of smaller catastrophic episodes in addition to the five major extinction events that have been so well documented (Courtillot and McClinton 1999; Hallam 2005; Hazen 2013; Knoll 2003). Since our own planet has an even more "angry" history than Darwin realized in 1859, it would seem that the cause and effect relationships with respect to evolution may be far more dependent on chance factors ("luck of the draw") than even Darwin could have imagined. Life on other worlds, if it exists, could possibly be different in terms of its physiology and chemistry from that found on our own planet since it would probably be the co-evolutionary product of a different physical environment combined with a different set of chance physical or biological events (Jones 2008). Not until (or if) mankind is able, sometime in the future, to begin sending sophisticated robotic space probes or teams of real living neuroscientists and biologists to far distant inhabited exoplanets (or the inhabited moons of some exoplanets) in the universe, will we be able to confirm or disconfirm how intelligent life forms may or may not evolve on other worlds (Coustenis and Blanc 2012; Schulze-Makuch 2013; Morris 2003; Ward 2005).

However, the author would like to emphasize that he, along with many other currently Earthbound scientists, believe we will not need to wait until mankind is able to send real live scientists to far away strange worlds to begin unraveling how different kinds of life forms are able to evolve in and adjust to different kinds of environments. Today's Earth-bound evolutionary paleontologists, with all their sophisticated new tools for collecting and investigating the fossilized remains of earlier life forms on our planet are presently very capable of making a major contribution to this research effort (Knoll 2003; Knoll et al. 2012). These people are well aware that they already have access to a large number of excellent examples of how life responds over time to changing environments. *In its 4.5+ billion year history, our own planet has done an amazing job of repeatedly imitating the hostile environments of many different kinds of alien worlds.* All our scientists need to do is put on their boots, hats, or suntan lotions (or knee pads and oxygen masks) and "follow the fossils" to investigate how this happened on our own planet.

The long and traumatic evolution of the Earth has presented our scientists with amazing amounts of potentially valuable raw data on how life is capable of adjusting to dramatic changes in planetary environments. Our planet may not be as "rare" a phenomenon in the universe as we now believe. Our planet was not always a warm comfortable blue water world with lots of refreshing oxygen to breathe. On many occasions in the past our Earth was an extremely hot and hostile world filled with poisonous CO_2 and other kinds of toxic gases that was thought to be not life friendly, at least for many of man's closest oxygen breathing relatives. In earlier years, our planet was more like that of many of the hostile alien worlds that our planet hunters are now reporting to be out there (Chela-Flores 2013; Grenfell et al. 2013; de Vera and Sackbach 2013; Schulze-Makuch 2013). In spite of this horrible debut our planet quickly and unexpectedly became quite life friendly, at least from mankind's point of view (Kasting 2010; Lunine 1999). Life began popping up everywhere as soon as our hot and molten planet cooled down sufficiently to allow the first water to exist on its surface. And then the battle for life's survival began in earnest on our planet! During the next few billion years, right on up to the present time, our planet repeatedly tried to rid itself of life by going totally hostile to it.

Our planet has gone through at least five major extinction events involving extreme outbursts of volcanic activity and/or catastrophic collisions with comets or asteroids from the most recent such event about 65 million years ago (which launched the extinction of the dinosaurs, see Fastovsky and Weishampel 2012) to as long ago as about 440 million years ago, when the Earth might have been hit by intense radiation from a distant exploding star (Melott and Lieberman 2013). Although the animal death tolls (both land and ocean) were estimated to have been at least 25–50 % or more in these earlier extinction events, the most extreme event (which resulted from a virtual worldwide outbreak of severe mantle plume volcanic eruptions) was the so-called "Great Dying" extinction event at the end of the Permian Eon about 250 million years ago in which over 90 % of all life was extinguished (Erwin 2006).

In addition to experiencing occasional disasters from asteroid strikes or severe volcanic activity, our planet also has a long history of periodically experiencing extreme weather-related disasters. The Earth has a long history of periodically warming up and cooling down in response to the gravitational influences of Jupiter and Saturn occasionally combining with the gravitational pull of our moon to produce slow changes in the direction and degree of our planet's orbital tilt. The most severe such climatic disaster occurred about 650 million years ago when the Earth experienced a total freeze down (the "snowball Earth event") in which the planet may have been covered by glaciers all the way from the north and south poles to the equator (Walker 2003). This "big freeze", which different scientists estimate may have lasted anywhere from 6 million years to as long as 10 or 12 million years, is thought to have destroyed huge numbers of species of single-cell ocean life that opened the door to the subsequent rise of multi-cellular life forms.[1] Scientists believe that the Earth has probably experienced at least four such "ice age" events in its history, although none were as severe as the one that occurred 650 million

---

[1] Experts in the field, while not agreeing on the exact numbers, have estimated that the total number of living species (whether single-celled or multi-celled) that currently exist on Earth is somewhere between 10 and 100 million. More sobering is the estimate that of all living species that ever lived on Earth, 95–99 % were destroyed when their home planet underwent geological changes that they could not cope with.

years ago. For at least the last 10,000 years, Earth has been in an inter-glacial period, and the next ice age is due to probably begin in a few thousand years unless of course, global warming intervenes in some yet unknown but possibly devastating manner that currently has some of our best Earth scientists very concerned. The residents of other rocky exoplanets in other stellar systems might also have to deal with similar re-occurring global temperature regulating problems!

Thus, it appears that life's best friend as well as its worst enemy is the environment in which it lives. Thus, the details of how it is that life is so tenacious (stubborn?) to any and all attempts by its home environment to destroy it is all around us (and especially below our feet) and is readily assessable to those of us who do not mind sweating, digging, and climbing down into deep, dark, wet, and hot caves, or large frozen icebergs, or the highest mountains and deepest oceans to search for it. The planet Earth we see today is totally different from the planet Earth of past years (eons). Although the author would never endorse it, the idea that, when it comes to the co-evolution of life and its environment, "that which does not destroy you will make you better" may have some merit to it. While science may not be a religious pursuit, it could be interpreted by emotional predatory creatures like us as having a definite spiritual aspect to it.

Not only did the Earth itself undergo a huge number of dramatic geophysical or climatic changes in its long evolutionary history which have provided raw data (opportunities) for our geologists and paleontologists to determine what kinds of biological adjustments did or did not work best in particular environments, the evolution of Earth's so-called extremophile life forms has also given our life scientists a collection of actual living specimens with which to perform additional "living experiments" to determine what kinds of specific chemical or physiological changes could work best to ensure the survival of living organisms in particular past, current, or even future hostile environments. Thus, in addition to not having to wait for rocket science to allow them to perform "on-site" investigations of how life adjusts to different physical environments, our scientists can also go into their own backyards or into their laboratories to determine how our extremophile relatives are able to survive in their own unfriendly neighborhoods. Astrobiologists are now finding evidence that many of the survival techniques used by extremophile species are not the result of complex new protective genetic mechanisms being added to the organism's basic biological structure but are frequently the result of simple modifications (dare I say "ingenuous"?) or alterations of existing physiological mechanisms.[2] For example, microbes that live deep inside frozen icebergs, instead of inventing new complicated genetic or metabolic processes to protect them from the cold, simply produce their own special "anti-freeze" chemicals. Likewise, organisms that might fall victim to being damaged or destroyed by exposure to excessive

---

[2] The observation that many of Earth's species select the easiest solutions for their survival in "extreme" environments is further proof that evolution tends to choose the least complicated means or solutions whenever a species needs some kind of evolutionary modification in order to cope with its changing environment.

ionizing radiation also appear to be capable of building and storing special protective enzymes (proteins) that they can use to make repairs to their damaged genes, when needed.

Since the author has now reached the "bottom of the barrel" with respect to his own knowledge of these profound issues of life and evolution, I will devote the remainder of this chapter to summarizing and expounding on a number of my own ideas and thoughts on these issues, as well as those of other respected, but definitely Earthbound scientists or thinkers in the Earth, life, and behavioral sciences. Unfortunately, the author will now have to run the risk of having some readers totally abandon their reading of the present book by making the profound statement that.... **except for an occasional "Rare Earth"** (see reference by Ward and Brownlee 2004) **out there, the smartest of our future astrobiologists may be unable to recognize the inhabitants of many exoplanets as being what they are, i.e., living or biological entities!** The reason for this is simply that, in the last few years, our space scientists have started telling us that it now appears that, of the multitude of exoplanets that are being discovered, only a relatively small proportion may be even close to being a twin of our own home planet (*at least at Earth's present stage of evolutionary development!*) with respect to all their physical and chemical features, including size, location, etc (Louis and Minelli 2012). Many exoplanets appear to be quite different from our Earth and, in many cases, even different from our own distant giant gas and ice planets (e.g., Jupiter, Saturn, Uranus, and Neptune), or our smaller sun hugging hot neighbor Mercury. The additional finding that some other planetary systems may host huge planets (super-Jupiters) that orbit so close to their home stars that they are incredibly hot and probably totally hostile to life (as we know it) has now convinced some planet hunters that the specific manner in which planetary systems are formed may also differ across the universe.

Also, starting at the end of the twentieth century, our biologists started telling us that life on our own planet is amazingly more diverse and complex, plus more resilient and adaptable with respect to its ability to survive in environments all over our world (plus above and deep inside our Earth) that would instantly kill any so-called "normal" forms of Earth life. Therefore, if life that is based on the carbon atom and water, like it is on Earth, is the only type of life that can occur in our universe, then "intelligent" extraterrestrial life forms that are even remotely similar to us in terms of their physiologies, chemistries, and behavioral characteristics, are bound to be relatively rare on other worlds. And, furthermore, the recent discovery of extremophiles that manage to live all over our world in the most hostile environments, now suggests that life, on our planet and perhaps elsewhere in the universe, may be even more tenacious than previously thought possible and can easily evolve or adapt to even the most hostile environments. Many life scientists now believe that such "tough" forms of life might not need to be dependent entirely on the carbon atom and water as it is here on Earth, but could be the product of completely different chemistries that co-evolved in conjunction with completely different kinds of environments (Ward and Bennett 2008).[3]

Our life scientists, and especially our biochemistry experts, now believe that the reason that our specific type of chemistry, which is based on the carbon atom and water, works so well to support life on our planet is that the structural and chemical properties of carbon, and also water, just happens, by chance (or otherwise), to work best in our kind of environment. Carbon is able to form all kinds of large complex molecules (i.e., "macromolecules" that may contain tens or hundreds of separate atoms, or even several millions in the case of the largest organic molecules we know of, i.e.; the nucleic acids RNA and DNA) that are needed to build life, and water is able to easily support the chemical reactions that our type of life depends on. **Thus, carbon and water work so well for us** *simply because our environment allows these things to happen*. The chemical makeup and pressure (density) of our atmosphere, plus our surface temperature, etc. are all just right to allow carbon and water to do their "jobs". It is also, of course, helpful that both carbon and water are readily available on our planet. Our carbon and water chemistries would probably not presently work on other planets in our solar system such as present day Venus, which is currently far too hot and dry, contains the wrong gases in its atmosphere (e.g., lots of carbon dioxide and sulfuric acid), and has an atmospheric pressure that is 90 times greater than ours. Our type of carbon/water chemistry also would not *today* work on one of the moons of Saturn (i.e., Titan) where things are presently far too cold (water is frozen solid on the surface but perhaps not so in slightly warmer underground oceans that NASA's Cassini spacecraft recently discovered) and the atmospheric chemistries are totally wrong (the planet smells more like a stinky oil refinery than a flower garden).[4] However, since carbon is obviously readily available on Titan, if we were able to trade liquid methane for water as the biologic solvent on Titan, some other kinds of carbon-based life might be able to evolve (Lunine and Butler 2007). It is possible that the fact that Titan appears to host volcanic activity which supports a substantial gaseous methane atmosphere might mean that these carbon-based life forms might even breathe hydrogen since oxygen is not available. It is interesting that growing numbers of our scientists now believe that, because our young sun (between 3 and 4 billion years ago) was colder than it is today, the environmental conditions on early Venus might have also been wetter

---

[3] Many scientists, believe that if you want to know what living creatures might look like on other exoplanets in which the environments are very different from our own home planet Earth, you must take into account the type of environment the creatures live in. **The physiological morphology (both internal and external) of alien life forms on other exoplanets is most likely dependent on and determined by the specific nature of the environmental conditions that they must adjust to in order to survive.**

[4] However, the author and many other scientists now believe the incredibly slow warming of our sun (plus that of most stars) in their normal lifespans may totally change this scenario in a few billion years. Many stars may exhibit an unbelievably slow evolutionary transition that might allow "warm-blooded" carbon-based life to eventually pop up over time in more distant colder regions of their planetary systems.

and cooler making this planet, at least for a short period, truly a twin of today's Earth and capable of supporting carbon/water forms of life not unlike ours (see Fig. 1.16). And, the red planet Mars, while today very dry, cold, and harboring a thin atmosphere might also, 3 or 4 billion years ago, been much closer to being a twin of Earth than it is today (see Fig. 1.17a, b). At that time, Mars may have had a thicker atmosphere (larger greenhouse effect) plus a hotter interior which supported the presence of extensive liquid water (small oceans or lakes) for its surface as well as volcanic gases (e.g., methane) for its atmosphere. Important new evidence has been accumulated in recent years by our space scientists (at both NASA and ESA) that strongly suggests that early Mars might have, at least for a short time, actually hosted such life friendly surface conditions (Jakosky et al. 2007). (**Hopefully, NASA's new _Curiosity Rover_, which touched down on Mars on August 6, 2012, will shortly confirm this for us**. See Fig. 4.2a, b.) And many scientists now believe that this ancient Martian life has not gone away but has managed to adapt itself to today's much more harsh and drier surface conditions by retreating into wet and warmer underground locations.

Our life scientists have also known for some time that carbon may not be the only atomic element that is capable of forming the complex macromolecules that are needed to support life. While many atomic elements are totally unable to electrically bond with other atoms to form molecules, there are other atomic elements that might be able to do so, _if they reside in environments that are just right to allow it to happen._ Carbon, in our earthly environment, is the most prolific of our 92+ atomic elements in being able to do this and can easily form stable life friendly molecular bonds (long chains as well as side chains and rings) with as many as four other atoms at once. The atomic element silicon is another type of atom that has many chemical similarities to carbon and, on Earth, can also form molecular bonds with as many as four other elements. Silicon does not, however, tend to form long chains containing many different atoms which is thought to be necessary for the formation of large complex molecules that are critical for life (at least as we think we know it on Earth). In Earth's environment, the bonds that silicon can make tend to be unstable. Some of these bonds are too strong to be easily broken when some kind of life friendly chemical interaction is needed, and also sometimes too weak to prevent molecules from being accidentally broken apart when they should not be. Another problem with silicon, at least here on Earth, is its extreme affinity to form bonds with oxygen. When carbon-based organisms on Earth breathe and exhale, they toss away carbon dioxide (CO_2), which is a gas. Silicon-based creatures might exhale silicon dioxide (SiO_2), which is sand and not a gas. A sneezing silicon-based life form might be able to earn a living as a living sand blaster!

The reason that biochemists have traditionally had problems discovering what different atomic elements might or might not be able to do with respect to forming complex life friendly molecules is that virtually all of the research that has been performed was conducted on Earth in air-conditioned or heated laboratories under sea-level atmospheric pressures. Other kinds of atomic elements might be able to perform their unique chemical tricks to support other kinds of life if allowed to do

Fig. 4.2 On August 6, 2012, NASA and JPL landed the new *Curiosity Rover* on Mars to begin mankind's long awaited renewed search for evidence of life on Mars. (**a**) Shows an artist's drawing depicting how the Curiosity Rover was lowered to the Mars surface by special tethers (ropes) from another rocket ship that hovered above the Mar's surface. (**b**) Shows another artist's drawing of the lander as it gets ready to begin its search for life on Mars. *Mankind is now once again seriously involved in conducting **on-site searches** for life on other planets or moons in our own solar system!* (image credits: NASA/JPL)

so in other environments that we "carbon-chauvinists" would likely label as being extreme or hostile. Although not investigated as thoroughly as silicon, some scientists have suggested that some other atomic elements such as boron, nitrogen, and phosphorus might be viable substitutes to the carbon atom in other kinds of non-Earth environments. Some scientists have suggested that other kinds of non-water solvents might also work in some other environments. Liquid ammonia, methane, and hydrogen fluoride have been most often mentioned, but other chemicals such as methanol, hydrogen sulfide, hydrogen chloride, or even extremely cold liquid nitrogen, have also been suggested. Thus, if science's recent discovery of extremophiles living all over our own planet does turn out to be an accurate indicator of the extreme flexibility of life in the universe, then future generations of astrobiologists may eventually begin finding really strange forms of extraterrestrial life that may not be easy to recognize as being other life forms. So, from the viewpoint of some of our extraterrestrial neighbors out there, it may turn out that *it is us and not them* that are the real strange creatures that inhabit so-called extreme hostile environments. Our form of life has even found a way to survive in an environment that is filled by one of the most chemically reactive (and biologically toxic) elements known (at least to man), i.e. oxygen. It does seem that, in the future world of our astrobiologists, "beauty *really* may be in the eyes (or other kinds of sensory organs) of the beholder"!

Therefore, human-like creatures like us may truly be the oddballs in our vast life-friendly universe and my earlier description of how life evolved on our planet may not be representative of how life might evolve on worlds with totally different environments. The noted Harvard biologist Stephen Jay Gould (2001), therefore, may have been totally on-target when he predicted that if the entire history of life on Earth could be recorded on a cassette tape (a very big and long one, with ultra-HD capacity, of course) and then, following some kind magical event in which the entire history of life on Earth could be repeated a second time, and again recorded on a cassette tape, virtually everything on the second recording would be changed from what it was on the original recording. Dr. Gould, along with many other evolutionary scientists, obviously believe evolution very seldom replicates itself. In contrast to Dr. Gould's interesting cassette tape analogy, the author long ago developed his own personal rendition of this "failure to replicate" idea which he inflicted on his own students and would like to now share with his readers—"*Very different kinds of life would likely evolve on our own home planet earth if we could somehow go back (in some kind of time machine) to our world about 4.5 billion years ago before it had finished accreting into a planet, wipe out any prebiotic organic molecules that might have already managed to hitch rides inside meteors or comets and land on our new planet's surface (or chain themselves together in some hot hydrothermal vent), and reset the new planet's evolutionary clock to completely repeat every physiological and geological event a second time exactly as it occurred the first time around*". Therefore, the distinguished evolutionary paleobiologist Simon Conway Morris (2003) may have been totally "right on" when he made the following statement in his recent book entitled **Life's Solution: Inevitable Humans in a Lonely Universe**:

> *The history of life is littered with accidents: any twist or turn may lead to a completely different world. The evidence demonstrating life's almost eerie ability to navigate to the correct solution is now being challenged. Eyes, brains, tools, even culture are all very much in the cards. So, if these are all evolutionary certainties, where are our counterparts across the galaxy? The tape of life can run only on a suitable planet, and it seems such earth-like planets may be much rarer than is hoped. Inevitable humans, yes, but in a lonely universe.*

So, while mutations, evolution, and natural selection are mostly pure random events, the way organisms adjust to their environments may not always be quite so random. The kinds (and order) of random genetic mutations which would pop up in such a hypothetical second round of evolution on our planet or any other planet would likely be very different from the first go around, but how they (the new traits) might interact (co-evolve) with the concurrent geological and climatic changes might or might not be so different. Evolutionary biologists tell us that, in a very real sense, the process of evolution does quite frequently utilize similar solutions to allow organisms to adjust to similar changing environmental conditions (in a process scientists call *convergent evolution*, see Futuyma 2005). The process of adapting to change is not always totally random. In the history of life on Earth, wings have been re-invented many times (in different physical formats, of course) to allow different species (insects, birds, flying squirrels, flying fish, airline pilots, etc.) to be able to fly through the air. And the importance of being able to see is so important to life on Earth that eyeballs were re-invented at least five separate times in life's history. And the eyeball design that nature provided man is not even the most efficient and sensitive version. Birds have far better visual acuity that we do.

Most scientists today believe that the process of biological evolution is extremely practical in the sense that evolution tends to use the simplest and least biologically challenging means available in a given environment to solve the survival issues of living organisms. Therefore, it is likely that, in a hypothetical replication of the evolutionary process in a particular environment (e.g., Earth), the animals, plants, and other life forms that would arise in a second evolutionary "go around" would not be clones of the original species, but might look very similar to the original species since they would be the evolutionary products of a similar or almost identical environment. Any fish-like creatures in the second group might have something very similar to fins that would allow them to swim, and any flying creatures would have something comparable to the wings of their pre-replication cohorts. It is likely that when we start discovering twin Earth type exoplanets, we will find life forms that look vaguely similar to our own life forms here on Earth. Something equivalent to our own legs, fins, wings, and other forms of jointed and "muscle"-like supported appendages, plus eyes, etc., could be relatively common on such Earth-like planets. This author, along with many professional astronomers and space scientists, therefore, now believe that life could possibly develop on some of those totally non-Earthlike planets that the Kepler space telescope people are finding (Louis and Minelli 2012). Whether or not there are any "universal laws of evolution", it appears that the evolution of life is quite frequently nature's simplest response to the specific environmental conditions that exists on any planet. Life on other twin Earths would, therefore, probably have similar external (and internal) morphological characteristics to life on Earth, while life forms on exoplanets that are radically different (in terms of size, distance from their home star, makeup of their atmospheres, etc.) would look radically different from Earth's life forms, and might even be unrecognizable to our astronauts as being living creatures.

Finally, I would like to briefly discuss another very important concept that, unfortunately, has been a frequent "thorn in the side" to scientists as they attempt to unravel the mysteries of how evolution works. If one looks at many of the different "family trees" that have been a common graphic feature of textbooks on evolution over the years, it is easy to draw the conclusion that the evolutionary process involves a step-like series of extremely slow advances over time in, for lack of a better term, what we might call complexity (or "efficiency"?). Sometime between approximately 3.5 and 4 billion years ago, the first life forms popped up on our still very hot and hostile planet. These simple single-cell organisms (called prokaryotes) were the only life forms on Earth for at least the first three billion years following the formation of our planet (Schopf 2002). Then, thanks in part to the advent of an oxygen dominated atmosphere that allowed more efficient metabolisms to develop, these single-celled organisms were joined by another group of larger and even more complex single-cell creatures that paleontologists call eukaryotes. And, a little over 1 billion years ago, these complex single-cell organisms started linking themselves together to form larger and even more complex multi-cellular life forms (Copley and Summons 2012). And then, in what we might label as a mere "eye blink" ago in life's history, this so-called "complexity race" suddenly exploded and even larger and definitely more complex plants and animals started popping up everywhere. And, in just the last few million years, the primates evolved and, just a little over 200,000 years ago, one branch of the primate lineage led to us modern humans (Haines et al. 1998; Haines and Chambers 2005).

However, the idea that evolution is inherently designed to allow life to advance from the simple to the complex may not be necessarily accurate but could be more an epiphenomenon that arises from a mix of multitudes of more simple, but real, environmental and genetic effects. For many years, hundreds of dedicated scientists (paleontologists, geologists, biologists, etc.) have, on hill sides, inside caves, and especially in laboratories, been trying to find evidence that this apparent march from the simple to the complex is indeed some kind of inherent goal or mandate of the biological evolutionary process. This concept, nevertheless, is now believed by many paleontologists to be a product of man's own anthropocentric mindset (a long-winded word that the dictionary defines as "considering man or his values as the central goal of the universe"). The evolutionary family trees that textbook illustrators draw only show the branching points where one species turns or changes into another more "advanced" species. They never show that those species that branched into another species did not go away but themselves kept on evolving. Evidence that this upward trend towards more and more complex (or even intelligent) life forms is not a totally real or even mandated phenomenon can be seen in the evolutionary record itself. Whales and porpoises evolved from ancestors that once were walking land creatures. The trading in of legs for fins does not seem to be any kind of evolutionary advance (unless you think like whales) although it is a good example that evolution can sometimes reverse itself.

One of the major consequences of all these exciting new discoveries of exoplanets and extremophiles is that many of our life scientists have now been forced to admit that their ideas of "what life is" have been shoved all the way back to the dark ages. At the end of the twentieth century, most of our life scientists believed that coming up with a viable definition of life was "possible but very difficult". Today, many of these same people, plus their students, have "thrown in the towel" and now believe that a definition of life may be virtually impossible. Many astrobiologists now speculate that life (or whatever "life" is) may be widespread and even rampant in the universe, and may exist in all sorts of weird environments and be constructed in many different formats (chemistries). Life "as we know it" on Earth may be only one of many different forms of life as we "*do not* know it". The recent sudden and dramatic rise of digital computers in our daily lives has now triggered some of us to even begin believing that some intelligent life forms in our universe may turn out to be mechanical machines invented by carbon-based (or silicon-based) biological entities like us, rather than being the product of biological evolution. A growing number of serious scientists are now beginning to believe that this transition from intelligent biological nervous systems to intelligent mechanical systems may have already occurred elsewhere in our universe (Shostak and Barnett 2003; Shostak 2009; Skurzynski 2008). Rather than looking for "little green persons" on other worlds, we might consider searching for intelligent machines that are capable of thinking, replicating, and even evolving that biological life forms may have invented and unleashed millions or billions of years earlier. In the next chapter, I will turn my attention to describing what I along with many very competent and brave scientists think may have already transpired on other worlds in our universe with respect to this bizarre transition from biological to machine-based intelligence, and how this may affect us in the not so distant future.

Chapter 5
Are Biological Nervous Systems Just the First Step in the Rise of Intelligence in the Universe?

In 1950, any book that might have had a chapter with this unusual title would have been immediately declared as being the ramblings of a deranged author or an excerpt from a really bad science fiction novel. However, in the last few years, our scientists have now made a number of absolutely incredible and totally unexpected breakthroughs in several different fields of science that now makes a chapter title like this, while still strange for many of us, perhaps not so farfetched for some others among us. Our life scientists now tell us that the first living creatures were actually able to evolve on Earth circa 4 billion years ago almost as soon as our hostile world had cooled down sufficiently to allow water to exist on its surface. And even more startling is the discovery that, even today, the descendants of these first life forms (which our scientists now label as extremophiles or "lovers of extreme environments") that can tolerate all kinds of hostile environments that would kill all the rest of us are still living and thriving anywhere on, under, or above our world that they choose as long as they have access to carbon atoms and water. Thus, life on our planet may be complex but it is definitely not always fragile, and it may be incredibly more adaptable than any of us would have believed possible just a few short years ago. And, since 1995 our astronomers have started telling us that other planetary systems may be common in our universe, and many of these exoplanets are not twins of Earth that might support carbon-based specimens like you and me but might be homes for life forms that are more similar to some of our own extremophile relatives.

Thus, it now appears that this entity we call "life" may not be as complex, mystical, or magical as many of us used to believe. It might actually be able to thrive on many other worlds in our universe, and even take on many different biological/chemical formats as well as possibly some kinds of non-biological forms. The idea that some intelligent living things out there may be some kind of non-biological machines rather than "blood-and-gut" products of something akin to biological evolution has now suddenly emerged as a real non-fictional possibility thanks to mankind's recent invention of something called the *computer*. In the next few pages, I will try to describe what I and many quite competent and hopefully

© Springer International Publishing Switzerland 2015

J.L. Cranford, *Astrobiological Neurosystems*, Astronomers' Universe,

DOI 10.1007/978-3-319-10419-5_5

sane people now believe is a scientific revolution that is exploding all around us. Our scientists now appear to be well on the road to creating a society in which some form of non-biological or artificial intelligence may very well dominate our future lifestyles. Is this good or bad, and has it already occurred elsewhere in our vast universe?

From Biological Intelligence to Computing Machines

If the evolution of intelligent biological forms of life on our planet is representative of how such things typically happen on other worlds in the universe, then the process is definitely incredibly slow. While it took over 4 billion years for the first humans to evolve, the time it took for mankind to develop a technological society was, in terms of cosmic events, quicker than the blink of an eye, or virtually instantaneous. Because of the rise of computer technology and rocket science in the last half of the twentieth century, many scientists now believe mankind is on the verge of discovering that this thing we loosely call "life" may be quite common in our vast and hostile universe. The first radio signals were successfully transmitted across the Atlantic Ocean shortly after 1900, and since 1960, Frank Drake in California and other scientists in research facilities all over the world have been busily using radio telescopes to listen for possible radio signals from intelligent life forms on other planets in our galaxy (Ekers et al. 2003). As I sit in front of my computer tonight writing this book, I am sincerely hoping that my brain will stay lucid enough to allow me to join my friends and fellow amateur astronomers in celebrating when, in possibly just a few short years, the news media finally breaks the news that our astrobiologists have confirmed the existence of a twin Earth somewhere out there and mankind's newest technology in the form of the James Webb space telescope (or its successor) has picked up evidence that the planet is not only warm and wet but harbors an atmosphere that appears to be life friendly for carbon-based creatures like us.

The evolution of biological intelligence on our planet was not only incredibly slow in occurring, it was, and continues to be, very inefficient. The smartest humans continually make mistakes when they attempt to operate their brains, and our speed of learning new things, figuring out solutions to everyday problems, etc., is unbelievably slow at times, and the so-called smartest of us are sometimes referred to as "absent minded professors". By the 1940s, mankind finally managed to develop something called the digital computer (Fig. 5.1).[1] At first, these "calculating machines" were slow, although faster than our brains, and frequently subject to

---

[1] Prior to the invention of the first "electronic" computers in the early 1940s, which relied on vacuum tubes to perform "super human fast" calculations, a few brave souls like Charles Babbage (in 1837) had invented much slower, as well as far more unreliable "mechanical" computers that involved taking a large array of different mechanical levers, pulleys, and other non-electrical controlled mechanical devices and cramming them into a space that filled up much of their

Fig. 5.1 The first digital computers were huge and vulnerable to breakdown ("crashes"). This photograph (First Argonne Computer, 1953) shows what a typical computer system looked like in the late1940s and early 1950s. An equivalent computer system that has the same amount of "computing power" would today (2014) fit in the palm of the reader's hand, except that it would be capable of processing information thousands of times faster, and with minimal errors! (image credit: Wikipedia)

error or breakdown in their normal designed functions. The earliest computers filled entire rooms, were extremely sensitive to any changes in temperature or humidity, and required groups of technicians (some even on roller skates) to be on continuous standby to be ready to quickly replace burned out vacuum tubes. Nevertheless, when they worked, i.e., performed their assigned calculations or other "intelligent"

---

inventor's living rooms. Charles Babbage actually dubbed his mechanical computer as the "Analytical Engine".

functions, these machines were remarkably accurate, and usually far more accurate than our brains.

Shortly after the first large, cumbersome, and relatively slow digital computers/ calculating machines were invented that relied on vacuum tubes, intelligent life on our planet suddenly entered into a whole new and totally unexpected dimension. In the late 1950s, two scientists (Jack Kirby and Robert Noyce) independently discovered that those gigantic computers that required huge rooms and truckloads of vacuum tubes to make them work could now be replaced by much smaller devices called *integrated circuits* that could be placed on tiny silicon chips that ranged from the size of the palm of a child's hand to the size of the tip of the reader's little finger. It is amazing that, over time, as their internal components got smaller and smaller, computers got even faster, and increasingly cheaper to build and operate. The modern computer age had literally exploded onto the world stage! Those large and unreliable computers that had filled entire rooms were now being replaced by tiny *microchips* and *microprocessors* that could be placed inside any electronic device that man needed to assist him in his daily life. Tiny computers began popping up everywhere—inside televisions, toasters, cars, washing machines, etc., plus most other household appliances, and any other mechanical or electronic devices that required their own internal "brains" to make them work faster and more reliably. Next came the slightly larger *mini-computers* that include what we today call the "personal" computers (or PCs) or the small "laptop" computers, or the even smaller handheld devices called IPads or IPods (and just recently, so-called handheld "smart" phones) that now allow us to do everything from computing to talking on video phones as well as playing computer games and "surfing" the internet. At the time of the writing of this book, those early computers that filled entire rooms have now been replaced by devices that can easily fit into the palm of the reader's hand (Fig. 5.2), and, most amazing, they are now very powerful and reliable, and can operate thousands of times faster than their predecessors of only a few short years ago. And, most importantly, they are now cheap with price tags lower than most household television sets. Humans are rapidly becoming "hooked" on computers for virtually everything they do in both their personal and professional lives.

Many very well respected and totally serious scientists today (Kurzweil 2000, 2006) even believe that man's brain is "*on its way out*" as the dominant domain (medium, substrate, or holder) of intelligent life on our planet, that is, if mankind does not do something really stupid to destroy itself first. While computers have, in a very short time, shrunk to the size of peas or much smaller, and expanded their simple operating speeds by factors of millions, billions, and, very soon, trillions or even mega-trillions, they remain basically calculating machines, and not true biological-like thinking or problem solving devices. We are still a long way from replicating "Hal", the menacing and devious super-computer with the bad attitude that our astronauts had to out-smart in the classic 1968 Hollywood movie entitled "*2001: A Space Odyssey*". As I described in *Chap. 2*, today's brightest neuroscientists are still very much puzzled by how the trillions of racing electrical neural impulses in our brains gets translated into our daily thoughts or even our occasional "brainstorms". Nevertheless, virtually all of today's neuroscientists now believe we

Fig. 5.2 In the late 1950s, those large unreliable vacuum tubes that had been the heart and soul of digital computers began to be replaced by very small silicon chips containing even smaller integrated circuits and microprocessors. Computers suddenly began to *really shrink in size*! Unfortunately, today's personal computers (PCs) still require large hardware accessories to accommodate human operators, including keyboards, monitors, plus mice (or is that "mouse(s)"?). This photograph shows one computer manufacturer's solution to this problem. Future computers that do not require manual interfacing with humans will be smaller than a grain of sand or much smaller (image credit: CompuLab, Ltd., Yoknaem Elite, Israel)

are on the fast track to solving this, *the absolute greatest of all of life's puzzles*, in the next few centuries, or even sooner. Of course, many scientists (including the author (Cranford 2011) are still very much concerned about the possibility that the *messenger will end up destroying itself*, since relying purely on our own biological *brain power*, appears to be far from the most effective and safest way to achieve this goal. For many years, our science fiction writers have had a monetary super-heyday in writing stories about how man and his future super-computers may end up destroying each other. In my last book (Cranford 2011), I made the bold suggestion (which is shared by many other scientists, both past and present) that the solution to the Fermi Paradox of *If there are so many intelligent civilizations out there in space, why have some not tried to contact us or otherwise reveal their presence to us in some manner?* may simply be that the *evolution* of *inefficient intelligent biological nervous systems may be highly correlated with the development of technologies that can lead to their owner's inadvertent self-destruction before they are able to contact other civilizations*.

Merger of Computer Science and Brain Science Quickly Revealing How Nervous Systems Work

While still strict speculation (or informed science fiction?) at the present time, many of our computer science experts do believe mankind will be able, in a century or two, to develop intelligent artificial systems (or robots) which, while not being biological, may be based on many of the same chemical and physiological (plus physical) principles that allow the traditional biological nervous systems to work. Many scientists now believe that the key to converting or translating computers into mechanical devices that can truly mimic or even expand on the functions of the human brain will be to unravel the nature of the biological processes that allow actual living brains to do their magic. The invention of digital computers has now begun to make that dream possible. Our neuroscientists have now developed new high speed digital "neural imaging technologies" that use fast super-computers to actually record and analyze the physiological activity of millions of individual neurons in the brain while human patients or volunteers are performing different kinds of motor or mental/cognitive tasks (Beaumont 2008; Bremner 2005; Brooks et al. 2002; National Academy of Sciences Colloquium 1998). These scientists are now beginning to seriously unravel or "model" (i.e., develop and run computer simulations that mimic real brain function) how our biological brains are able to translate the complex physiological processes that are constantly occurring in the trillions of neural pathways in our brains into the so-called higher level activities of our minds. Machine-based mind reading is not that far away, nor is machine-based assisted mental or cognitive activities. The advent of incredibly fast and powerful digital computer technology is quickly providing brain scientists the tools they need to "look inside" the human brain and determine exactly what is going on when we see, hear, think, problem solve, or perform complex mental or cognitive functions (NAS Colloquium entitled "Neuroimaging of human brain function", 1998). While human volunteers (or patients) are actually performing complex mental or motor tasks, multitudes of tiny sensory electrodes or other devices can be placed close to the head or on the scalp that allow computers to obtain measurements (or images) of how different brain structures are involved in perception, thought, and physical actions.

The first of these super sensitive neural imaging techniques, known as *magnetic resonance imaging* or *MRI*, first arrived on the scene in the 1970s as a direct result of the invention of digital computers. This neural imaging procedure, like its digital computer "parent" has been quickly becoming more powerful, reliable, and sensitive. The MRI allows scientists to obtain amazingly detailed three-dimensional "pictures" of the smallest brain structures without having to use potentially harmful X-rays (Fig. 5.3a, b). The only radiation involved with this procedure are magnetic and radio waves, both of which are totally safe to use. MRI involves the electrical realignment of hydrogen atoms using changing magnetic fields. With this technique, the volunteer's head is placed in a barrel-like structure, which contains a large magnet that completely surrounds the head. Hydrogen atoms are the

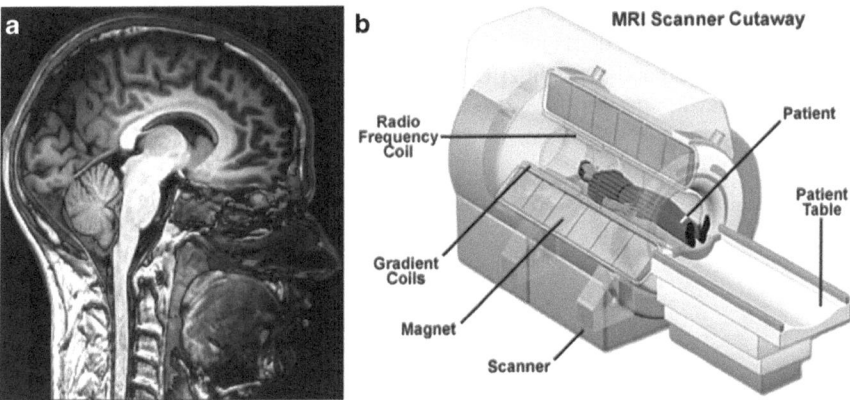

Fig. 5.3 (a) Shows a *magnetic resonance image (MRI)* recorded from the brain of a normal adult volunteer (image credit: Wikipedia Commons), while (b) shows an artist's schematic diagram of a modern Magnetic Resonance Imaging machine (image credit: National High Magnetic Field Laboratory, Los Alamos National Laboratory)

predominant ingredient of the water molecule, H_2O, and water is the most common component of most tissues, including brain tissue. The water content of the different tissues of the brain (nerve cells, myelin sheaths, axon and dendrite fibers, etc.) varies and, thus, the amount of water content (and hydrogen) varies. The surrounding bone (skull) contains even less water. When the magnet is turned on it causes the normally random alignment pattern of all of the hydrogen atoms in our brains to suddenly realign together in the same direction. When the magnet is turned off, each hydrogen atom "announces its presence" by emitting radio waves that can be picked up and recorded by special instruments. During the test session, the magnet is continually turned on and off to allow the development of a three-dimensional image that represents extremely small deviations in the hydrogen content in different regions of the brain and surrounding tissues. The turning on and off of the magnet can be noisy and some volunteers (or patients) have to wear acoustic earplugs to reduce the annoyance. Some volunteers are also claustrophobic and do not like being stuck in a small enclosed barrel-like space. In more recent years, the newer MRI systems have dumped the barrel concept and moved to "open space" systems (as well as much quieter models) that eliminate these problems.

The MRI procedure allows exquisitely detailed views of anatomical structure in the brain. However, while it is an excellent tool for identifying structural anomalies in the brain including tiny tumors, strokes, injuries, or other kinds of extremely small anatomical changes that may be interfering with normal brain activity, it is an event fixed at one point in time that provides no clues related to changes in ongoing brain function. This limitation, however, is quickly changing. Another computer based technique, known as *positron emission tomography (PET)* (*PET scans*) actually allows scientists to determine which specific regions of the brain are involved in some kind of behavioral or physiological task (i.e. are "working") by measuring which ones actually consume more oxygenated blood containing glucose (Fig. 5.4a, b). When a part of the brain becomes active (goes to work), it requires more energy to do its job. The brain area will need to take in more nutrition to fuel its increased activity. With PET scans, the neuroscientist injects glucose into a major artery (like the carotid artery). If a region of the brain becomes active the glucose will travel to that region to feed it. Before injecting the glucose into the

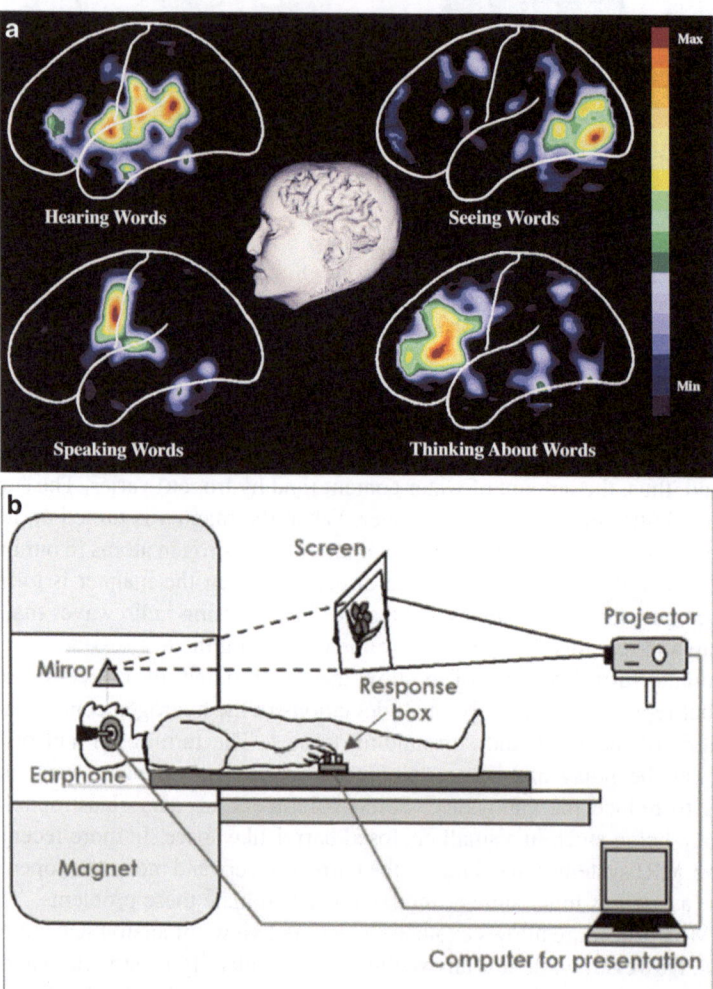

Fig. 5.4 (**a**) Depicts *positron emission tomography (PET)* scans performed on a normal volunteer as s/he was hearing, seeing, speaking, and thinking about real words (image credit: Marcus E. Raichle, M.D., Department of Radiology, Washington University School of Medicine, St. Louis, Missouri). (**b**) Shows how a PET scan actually records activity from brain cells as they are performing some kind of mental task (e.g., listening or thinking about words). The neurons that are actively involved in the thinking or listening task are "working harder" and taking in increased amounts of radioactively labeled glucose to feed themselves (image credit: Wikipedia Commons)

volunteer the scientist mixes it with a small portion of fluid that contains radioactive isotopes. This fluid serves as a radioactive tracer that constantly emits small (medically safe) amounts of gamma rays that can be picked up and recorded by the PET scan equipment. The PET scanner, thus, can trace where the labeled glucose goes and distinguish areas of the brain that are more or less busy at the time. If the patient is looking at pictures, the visual areas of the brain's cortex will "light up" indicating they are working; if sound is presented to the ears, the auditory cortical areas will light up. It is very interesting that patients with schizophrenia, who are exhibiting visual or auditory forms of hallucinations, will show increased activity in the visual or hearing areas of the brain the same way they would if they were seeing or hearing real objects. The brains of these patients are functioning (acting) as if they really are actually seeing or hearing something!

More recently, scientists have begun to develop ways of combining MRI and PET scans into a single test system that allows obtaining both functional and anatomical information on the same patients (or research volunteers). This exciting new tool is called *functional magnetic resonance imaging MRI* (fMRI). This technique promises to open the door for scientists to be able to see exactly where in the complex structure of the brain specific forms of neural function are occurring. The fMRI technique involves performing MRI testing at the same time that PET scans are being performed. The regions of the brain that require increased flows of the radioactively labeled glucose are literally plotted out on top of the detailed underlying anatomical structures that have been revealed by the MRI procedure (Fig. 5.5a–c).

Functional MRI is, at present, the most powerful computer-based technique we have for studying how brains work. However, current fMRI systems only tell us which areas of the brain become electrically activated when research subjects or patients are performing specific sensory, cognitive, or motor tasks. fMRI tells us where in the brain things are happening but does not tell us "what" is happening. Our neuroscientists tell us that, as far as we now know, all of our known forms of Earth-bound brains or nervous systems work by sending both electrical changes (nerve impulses) as well as chemical signals (via chemical neurotransmitters) from neuron to neuron. In recent years, teams of neuroscientists and computer scientists at the Massachusetts Institute of Technology and the California Institute of Technology have developed a new form of fMRI that can reveal not only which areas of the brain are activated electrically but also how chemical neural transmitters transmit messages between neurons in the brain. These scientists are studying how the neural transmitter called dopamine which is of particular interest since, of all the different kinds of chemical neural transmitters that the brain uses to transmit information between neurons, this one has an important role in motivation, reward, addiction, and several neurodegenerative diseases Thus; as computers get faster and more powerful, they are quickly allowing our neuroscientists to more accurately and thoroughly simulate and understand exactly how brains work. And since computers are quickly becoming far more reliable, accurate, and even faster than biological brains, in the not so distant future artificial intelligent machines will most likely make our present biological brains look like the truly primitive

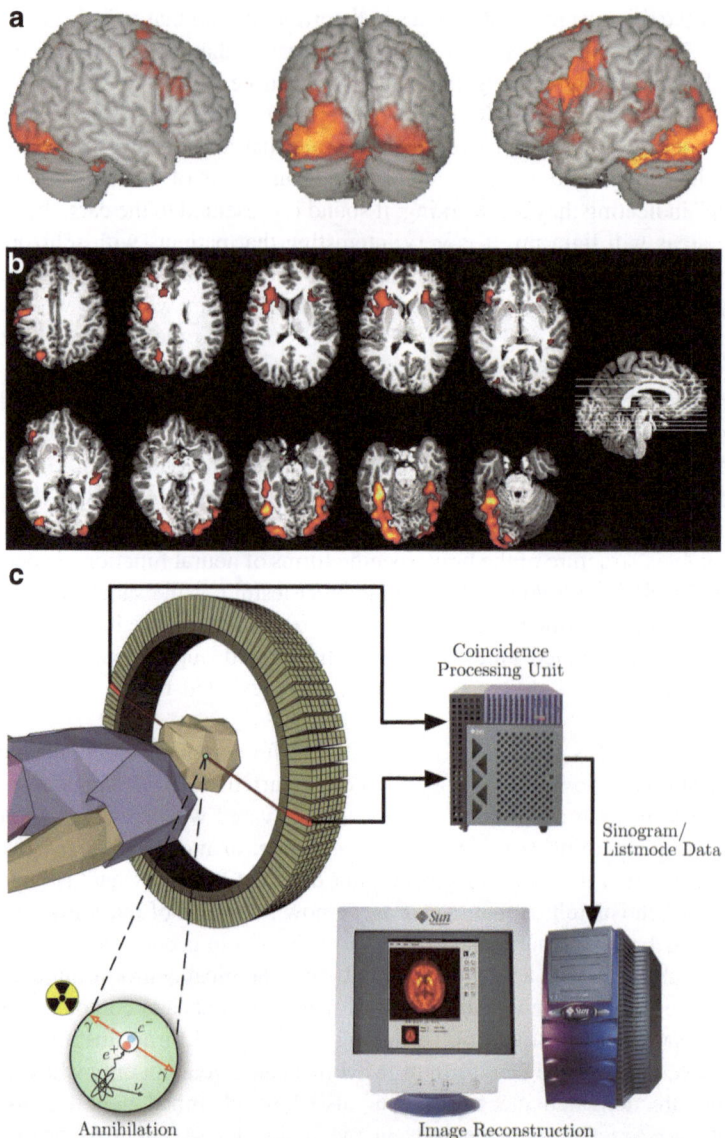

Fig. 5.5 (**a**) Shows an actual functional magnetic resonance image (fMRI) performed on a volunteer that was reading a list of words out loud. In order to view active regions at different levels below the surface of the brain, scientists record a series of images (called "slices") that, as shown in (**b**) which reveals activity at different depths or levels of the cortex (image credits: Thomas Thesen, DPhil, New York University). Finally, (**c**) shows a schematic diagram of a typical test system that is used by neuroscientists to obtain functional MRIs on research volunteers *at the same time* that they are performing some kind of behavioral task involving either looking at written words or listening to words being spoken by the tester (image credit: Lister Hill National Center for Biomedical Communications, U.S. National Library of Medicine, National Institutes of Health)

prototype biological control systems that they really are (Brooks et al. 2002; Kurzweil 2000, 2006).

The amazing fast pace of mankind's new "computer age" is, therefore, quickly providing our psychologists and brain scientists with the incredible ability to literally "look inside" the human brain and see exactly what is going on when we perform simple or even complex kinds of mental activity. Computerized "mind reading", whether good or bad for our society, is now just a few years away. And what really stimulates and excites our brave new breed of neuroscientists is the high probability that these new artificial brain systems will be able to work at incredible speeds (perhaps close to light speed) and will be almost totally accurate (Gardner 2003, 2007). Of course, if such super fast and accurate intelligent artificial machines (for want of a better word) do start walking, flying, or radiating out of our science laboratories in the next 100 or 200 years, what about us "mortals"? Will intelligent artificial systems eventually replace us slow-witted, clumsy, and error-prone intelligent life "prototypes"? Mankind will likely not be able to answer this question in any Earth-bound laboratories until well into the distant future, if life on our planet can survive that long. Nevertheless, the answer to this question may be waiting for us somewhere out there in space, and our astronauts may stumble upon it before the end of the present century or the not so distant future (Shostak 2009).

Could Intelligent Machines Become "Conscious" and "Self-Replicating" as Well as Capable of Some Form of "Self-Evolution"

On Earth, it has now only been a little over a hundred years since man stopped riding horses and pulling wagons and began contemplating rolling down the road or flying above the clouds in some kinds of noisy environmentally polluting machines. In spite of this, some of us have now seriously started thinking about escaping our planet and looking for other creatures like us in the universe. If the amazingly slow and inefficient process we call biological evolution is indeed the first step in the rapid evolution of technologies that allow truly intelligent (and reliable/accurate) systems to arise in the universe, what may our first space explorers find out there? Many of today's brightest scientists now believe that mankind's first encounter with extraterrestrials (ETs) will not be with real "blood and gut" products of some natural biological evolutionary process somewhere, but may be with either their technologically manufactured assistants or successors (Shostak 2009). This may, in fact, explain the Fermi Paradox. While our own Earth-bound computers are rapidly getting smarter and smarter, they have yet to show any signs of developing any "desire" to interact with us "socially". They are, and will likely continue to be for quite some time, only boring calculating machines that only do what their inventors tell them to do. When our SETI scientists send radio messages to the stars, the messages may be received but not "listened to". Therefore, since ET could be a machine and not a warm-blooded biological entity of some kind, she/he/it may not care whether or not we exist and will not bother to return our call. And lest we forget—"caring" and "bothering" along with many other aspects of the emotional or human side of our own biological brain function are the direct result of mankind's early extremely violent lifestyle. In order to stay alive and reproduce, our early ancestors had to kill and eat each other. Man and most other animals on Earth, are the direct products of such predatory histories (MacLean 1990). The evolution of the human brain was definitely dominated by the need to allow us to be able to quickly and accurately respond to threats from an extremely hostile and life-threatening environment. Man's invention of, and possible self-destruction by nuclear weapons or chemically altered biological weapons, is a direct result of our "natural" histories. Computers can, therefore, be "cold" non-emotional entities that do nothing but what they are told to do by their sometimes emotional programmers. Being free of all the "emotional garbage" that haunts the human brain allows computers to perform their jobs (calculations) really fast and really accurately. Unless their human owners program them to self-destruct (or mess up their construction somehow), they remain free of the possibility of destroying themselves (and us). Apparently, not so with their biological counterparts!

While our universe (or at least the part we think we can see) is estimated to be approximately 13.7 billion years old, carbon-based life has been present on Earth for slightly less than 4 billion years, and intelligent modern humans have been here probably no more than a few hundred thousand years. And, while human civilization started up a little less than 10,000 years ago (*when Earth's most recent "brief" inter-glacial climate change made it possible*), mankind's invention of the first digital computers occurred several years after the author was born, i.e., less than 50 years ago. And now, early in the new twenty-first century, man is beginning to take his first bold steps to leave our planet and begin looking for living creatures like us elsewhere in the universe. Most of our scientists today believe that, if intelligent civilizations are out there (or *still* out there), they would most likely, because of the age of our universe, be thousands, millions, or even billions of years ahead of us with respect to their technological skills.

Therefore, it is entirely possible that the first ETs our space explorers will encounter will not be intelligent biological life forms, or even sophisticated intelligent machines or computers, but far more advanced robotic machines that are capable of not only thinking and problem solving, but possibly also capable of self-replicating and even "self-evolving" that ancient dying carbon-based (or silicon-based, boron-based, phosphorus-based, etc.) life forms may have invented and unleashed many eons earlier into the universe to carry on their legacies as first suggested by Professor John von Neumann almost 60 years ago (von Neumann 2012). While some scientists have proposed the idea that, far in the future (or perhaps already in some remote parts of the universe) intelligent machines will develop the capacity for self-replication, the idea that machines could ever become capable of self-evolution seems a bit far-fetched for most of us to contemplate especially if you consider that normal biological evolution is supposed to be based on random gene mutations being linked with some kind of survival of the fittest concept. Still, once such futuristic intelligent machines become capable of self-replicating themselves, it would probably not require much more "effort" for such machines to also develop the technology they would need to be able to assess their current environmental surroundings and determine whether they might need to re-design, re-build, re-program, or re-tool their own internal operating and mechanical systems to increase their chances of survival (or make some kind of "improvement" in their present "mechanized" lifestyle). Such mechanical evolutionary changes would likely only require a short time (seconds, minutes, hours) rather than the millions or billions of years that normal biological evolution requires.

If the self-evolution of intelligent mechanical systems were to eventually become a part of mankind's futuristic world, it could provide a means of protecting life from being ravaged or destroyed by its immediate hostile environment. If this were to occur, we humans will have finally and totally freed ourselves (*maybe*) from all the horrible things that our own inherent predatory nature has inflicted on us for so many eons. (I say "*maybe*" simply because this futuristic science world may inadvertently come with a whole new set of "obstacles/impediments" that our descendents will be forced to deal with). However, at least for the foreseeable future (and probably far beyond), the natural evolution of biological intelligence (at least on planets like ours) will not only remain nature's primary means of building new life but will also continue to be so slow and inefficient that it will provide more than adequate time for many such systems to self-destruct or for our extremely complex and hostile (yet, paradoxically, life friendly) universe to destroy them. In the next chapter I will turn to a discussion of these somewhat more immediate and depressing ("emotion" stirring) possibilities.

Chapter 6
Just How Hostile Is Our Universe to the Development and Survival of Life Forms?

In the present book, in addition to taking advantage of my professional background as a neuroscientist to take on the difficult task of discussing how I believe alien nervous systems might be able to develop on other worlds in the universe, I also wanted to tackle the complex issues related to how difficult it might be for any possible life forms to survive in what we now believe is a universe that has frequently been characterized in the popular and scientific literature as either "life friendly" or, at other times, "unfriendly or hostile" to the existence of any known forms of life. This is, without any doubt, quite problematic to the story I am trying to present in the present book. What I would like to do is focus on how different kinds of life in different parts of the universe are affected by their own local environmental conditions. However, since I have knowledge of only one kind of life on only one kind of planet, I will have to limit my discussion to what our scientists currently believe they know about how our world may have, in the past, or could, in the future, host physical events that could support or impede the development of life right here under our own noses.[1]

Since 1995 our astronomers in their role as planet hunters have started finding evidence for the existence of thousands of exoplanets circling other stars in our galaxy. It now appears, however, that only a very small minority of these exoplanets may host environmental conditions that would allow our carbon-based form of Earth life to exist. Twin-Earth type planets appear to be quite rare out there.

---

[1] The author admits that our scientists frequent use of the terms "hostile" and "life friendly" in the context of how life and the universe interact with each other is an excellent example of how mankind's mental life (thoughts) are virtually a slave to our common language system that we acquired very early in life. Carl Sagan once said that the universe seems neither benign nor hostile, but merely "indifferent". That may sound "intellectually" more acceptable, but is it really? In Chap. 7 the author addresses the very real possibility that it may be the inherent illogical nature of our language lives that is "at the heart" of the "mind boggling" response that many of us (scientists as well as non-scientists) frequently experience when trying to understand how life and the universe works.

© Springer International Publishing Switzerland 2015
J.L. Cranford, *Astrobiological Neurosystems*, Astronomers' Universe,
DOI 10.1007/978-3-319-10419-5_6

However, our life scientists have now begun to discover that the entity we call life may be even more flexible, versatile, and capable of possibly surviving in some of the most extreme environments. In addition to exoplanets being everywhere, extremophile forms of life may also be relatively common. If this is correct, then it definitely makes the task of someone like me, who is quite familiar with only one form out of many possible alternative forms of life, very difficult. Thus, life may not be the extremely rare and fragile phenomenon that can only survive in extremely rigid life friendly conditions that our scientists, for so many years, assumed was the case. It is possible that life may widespread and may be able to accommodate itself to all kinds of extreme environmental conditions.

Therefore, if I lived in Las Vegas, Nevada, I would have no problems risking a large portion of my meager wealth in wagering that life is a common biological phenomenon that occurs on many other worlds in the universe. While more primitive or simple forms of life (e.g., single-cell or microbial organisms) are possibly quite common, as to the question of whether more intelligent or advanced/complex forms of life (e.g., multi-cellular, if "cells" are the "in thing" elsewhere) are also common, I might be hesitant to place anything more than a small bet. My reason for believing that simple life is more common than advanced life is based on two factors. First, since we live in a universe that is, in some sense, probably "hostile" to the development of any form of life, simple life forms that can arise more easily and more quickly undoubtedly have an advantage over advanced life forms which require far more time and resources from their environments to develop. However, another reason this author believes simple life is probably more common than advanced intelligent life is that I have far too much knowledge (based largely on my professional background in psychology and the brain sciences) as to how much more vulnerable intelligent creatures may be to self-destruction because of their predatory histories. If the manner in which intelligent life evolved on our small rocky world is representative of what must happen on other worlds in order to allow either similar or different forms of intelligent life to develop, such intelligent ETs must be able to very slowly (over many millions or, more likely, billions of years) navigate their way through a very long and extremely unfriendly obstacle course that is filled with huge numbers of physical or environmental pitfalls that can, at any time, completely destroy it or force it back into a much earlier and more primitive stage of existence. And, unfortunately, many scientists (MacLean 1990), including me, believe that, because an earlier predatory lifestyle may be one of the important keys to the development of intelligent nervous systems, the more intelligent the life forms become, the more vulnerable they may become to inadvertently destroying themselves with the technologies that their intelligent natures permit them to develop I will have far more to say about this possible "down side" to the rise of intelligent life on our own planet, and perhaps elsewhere in the universe, later in the present chapter.

Since our home planet Earth is the only place that we know exists that is literally chocked full of life in every nook and cranny (albeit, of only one kind which is based on the carbon atom and water), the author will be forced to use the Earth as the only example of what can go wrong to prevent life from developing, or to destroy it. Fortunately, our life and Earth scientists have, over the years, been able to collect a very large amount of evidence as to how life was able to happen here in spite of an untold number of catastrophic events that on numerous occasions (five we know of, but probably others that we do not yet know about) almost turned Earth back into a dead world. Therefore, in the first part of the present chapter, I will list and briefly describe the most common kinds of potential physical or environmental features of the universe that we know have been major threats to life on our own planet which might also destroy or prevent the development of either primitive microbial or advanced intelligent life on other exoplanets.

As far as scientists are presently aware, there are basically only three major kinds of dangers or threats to life in the universe that could occur. The *first* of these are *external dangers* from other objects or forces that are external to exoplanets, such as impacts from asteroids, comets, or even larger wandering objects, plus lethal doses of radiation from nearby exploding stars. The *second* kind are *internal threats* from within home planets such as sudden geological or climatic upheavals or catastrophes that could either destroy life entirely or damage it sufficiently to force it to return to a more primitive stage of development. And, finally, in the second half of this chapter, I will discuss a *third* kind of threat, which is the one that, as a retired professor of psychology and the brain sciences, concerns me the most and that is the real possibility that **many advanced extraterrestrial civilizations may be unable to control their own inherent primal aggressive/emotional natures sufficiently to prevent them from destroying themselves with the technologies that their intelligent natures have allowed them to develop**. The author will now describe each of these three forms of threats to the rise of any kind of life, and in particular intelligent life, which, ironically, might again help to account for the Fermi Paradox.

Most Common External Threats to Evolution and Survival of Intelligent Life on Planets

Since the 1970s, our life scientists have started finding amazing new evidence that indicates that life, while being complicated and frequently even fragile, may be "tougher than nails" when it comes to popping up in the most hostile locations and even surviving or adjusting to the most horrific environmental upheavals or catastrophes. In fact, mankind may be one of the more "wimpy" species living on our planet. Cockroaches, alligators, horseshoe crabs, and many other animal species have been here for millions of years and might be able to outlast mankind even if we do prove to be stupid enough to render our own environment totally unfriendly via nuclear wars or premature global warming. Man's predatory history could, at any moment, combine with his intelligent nature to be the source of his own final demise. Since the one example of intelligent life we find on Earth strongly suggests that intelligence requires at least millions and probably billions of years to be achieved, it would seem that the chances of any life forms surviving long enough in our incredibly hostile universe to be able to self-proclaim themselves as intelligent may be very small. In this chapter, I will try to describe the most common and well known of these threats to life. Additional threats will undoubtedly be discovered in the coming years as man (if he survives) gathers more information about what is happening out there in space.

Threats from collisions between planets and left-over debris from planetary system formation External threats to any life on habitable worlds are those catastrophic physical events that originate directly from or in the vicinity of other celestial objects in regions close enough to planets that they can have a dramatic destructive effect on life itself or the geological and climatic support systems that are critical for life. The good news is that, because of the vast distances between any habitable exoplanets and their closest celestial neighbors, all but the most profound threats or dangers would be located too far away to do anything more than put on a colorful show for any alien astronomers' cameras. With respect to external threats to the survival of life forms anywhere in the universe, the most common are impacts or collisions between habitable planets and other objects (e.g., meteors, asteroids, comets, or leftover debris from their stellar system's early formation process) that happen to share the same planetary system (Fig. 6.1a–c).[2] During the earliest stages of the formation of a new planetary system, such collisions are relatively common. On Earth, in the geological *Hadean Period* (or what our scientists commonly refer to as the *late heavy bombardment* period) that occurred between 4.5 and approximately 4 billion years ago, our planet was being pummeled by such huge objects every few hundred years (Gargaud et al. 2009; Hazen 2013; Knoll et al. 2012). In later years, e.g., from 1 or 2 billion years ago to the present, the frequency of these bombardments dramatically decreased and the size of the striking objects got smaller and smaller. The last major impact event that severely affected our own planet occurred about 65 million years ago which most scientists believe was involved in the extinction of the dinosaurs and the rise of mammals and eventually humans (Haines et al. 1998; Fastovsky and Weishampel 2012).

The threat of impacts from left over debris from the formation of planetary systems is, therefore, probably one of the most serious of the different external threats to life on any exoplanet. Early in the history of such planets, this threat is especially ominous since such collisions happen much more often and the impacting objects are much larger. This threat not only makes it very difficult for

---

[2] In our own planetary system, the primary credit for the removal of leftover debris (asteroids, comets, meteors, etc.) needs to be assigned to the presence of the giant gas planet Jupiter. Jupiter's huge gravitational field has acted as a magnet to make large threatening objects strike it rather than striking other smaller planetary neighbors like the Earth (see Fig. 6.1c). In addition to shielding our planet by stepping in to absorb the blows of menacing objects, Jupiter's strong gravitational field has also assisted us by slinging large amounts of such debris out of our solar system and away from us. Other planetary systems might or might not have such "body guards" to help protect their smaller exoplanets.

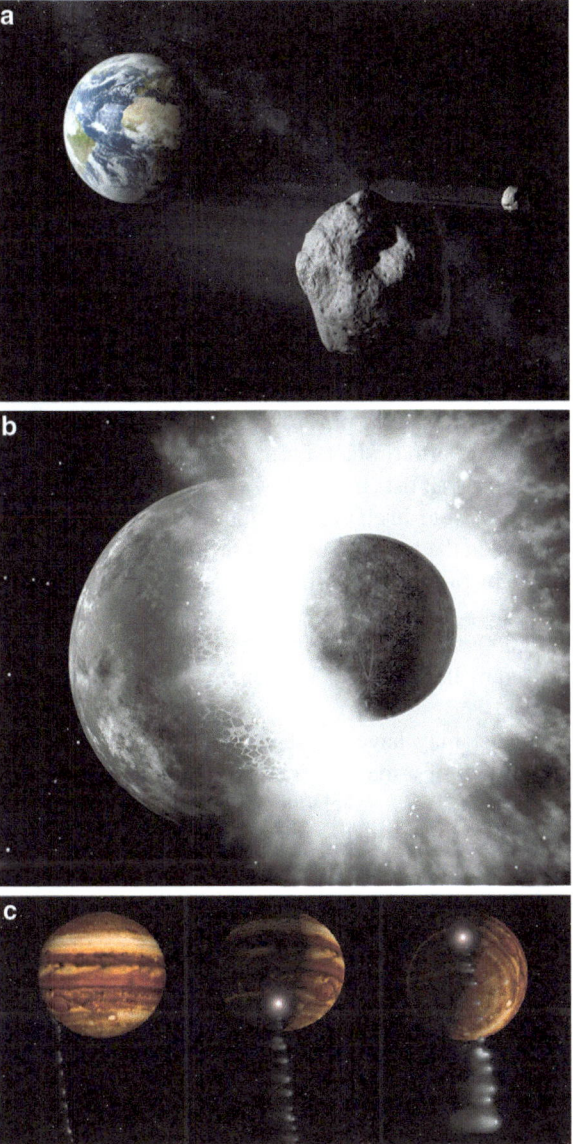

Fig. 6.1 Artists' drawings of (**a**) an extremely large "near Earth object" approaching and (**b**) colliding with the Earth. Astronomers do not believe there is any leftover debris in today's solar system that is large enough to inflict this amount of destruction, but a glancing blow from a large enough object (i.e., one of the many hundreds or thousands of smaller "proto-planets" that were still accreting or growing into our present pool of eight full size planets) like this could have been the way our moon was created circa 4.4 billion years ago. Inhabitants of other far away exoplanets might, however, need to worry about this possible disaster. However, (**c**) does show photographs of one very recent comet (Shoemaker-Levy comet) that struck the planet Jupiter in 1994 that might have, had it hit the Earth, produced widespread damage, including catastrophic loss of life worldwide. This comet, as it approached Jupiter, was torn apart into a series of smaller fragments by Jupiter's gravity. The largest fragments that hit Jupiter were 1.2 miles in diameter and were traveling at a speed of 37 miles/s. Thus, Jupiter's reputation as being our solar system's "protective

life to even get started, it can also threaten the evolution of more advanced forms of life. Since our geologists, paleontologists, and life scientists now believe they know a considerable amount about what happened to our own world when, about 65 million years ago, our planet experienced such a horrific disaster, the author will briefly describe the sequence of catastrophic events that we believe transpired when an asteroid (or possibly a comet), thought to have been somewhere between 6 and 10 miles in diameter, which is estimated to have been equal in volume to Mount Everest, slammed into the northern coast of the Yucatan peninsula in the Gulf of Mexico (Figs. 6.2 and 6.3). That this extinction event was primarily caused by an asteroid or comet impact and not some other kind of sudden extreme weather-related or geological event is provided by two pieces of evidence. First, scientists have identified the existence all over the world of a layer of sediment (that ranges from a half inch to well over three feet in thickness) at the same level below ground that contains ash from forest fires plus a special type of metal called *iridium* that, while extremely rare on Earth (in fact, far more rare than gold), is quite common in interstellar space. Most scientists now believe this rare metal could have only come from the exploding asteroid itself. In addition, the remnants of a 112 mile wide crater has been found on the northern coast of the Yucatan peninsula in Mexico that has been determined by radioactive dating to be approximately 65 million years old. This asteroid (or comet), which was moving at a speed 60 times that of a rifle bullet (approximately 45,000 miles/h) when it struck the Earth, may have produced the explosive energy equivalent of 100 million hydrogen bombs, which created a world-wide disaster.

The asteroid impact may have produced, over much of the North American continent, the equivalent of a magnitude 13 earthquake (the San Francisco earthquake of 1906 was a mere 7.8 level on the Richter scale used by geologists to measure the intensity of earthquake activity). Gigantic tsunami generated tidal surges (which normally result from sea bottom earthquakes, but can be the result of huge objects slamming directly into water-covered ocean areas) that could have been between 300 and 500 ft. in height or higher likely devastated vast regions of what is now the western, central, and eastern continental United States plus areas as far away as New Zealand and Antarctica. In addition to immediately destroying virtually all animal life within a radius of 5,000 miles, the impact threw huge amounts of hot and melted debris into the upper atmosphere that spread all around the world. As the extremely hot debris reentered the atmosphere it triggered massive forest fires all over the world. Worse yet, a vast cloud of dust and smoke particles filled the upper regions of the atmosphere world-wide that blocked the sun for 4–6 months plunging the entire planet into almost total darkness (except for the light from nearby burning forests). The lack of sunshine, in addition to producing near freezing temperatures, also virtually stopped all growth of new vegetation and

Fig. 6.1 (continued) shield" from impact events may have prevented mankind's early extinction (image credits: **a**—European Space Agency/P. Carril; **b**—NASA; **c**—NASA/JPL/Arizona State Univ.)

Fig. 6.2 While the asteroid that hit the Earth 65 million years ago was not big enough to destroy our world, it did produce widespread global destruction, including the demise of the dinosaurs. This artist's drawing shows what that asteroid impact might have looked like from high above the Gulf of Mexico (image credit: Don Davis/NASA)

Fig. 6.3 Shows a map of the location of the asteroid impact in the Gulf of Mexico that occurred just off the north coast of the Yucatan Peninsula in Mexico (image credit: NASA)

began severely disrupting the world's food chains. While the dust and smoke particles were largely removed from the atmosphere after 6 months, the sun did not start shining again for another two to three years.

The worst of the catastrophe, however, was not yet over! The asteroid had impacted in a part of the world that had a very large amount of sulfur-bearing minerals buried underground. The asteroid not only threw huge amounts of dust, smoke, and debris into the atmosphere but also huge amounts of rocks and soils containing sulfur. This sulfur immediately bonded chemically with the oxygen in the atmosphere to produce huge quantities of a very toxic type of gas called sulfur dioxide (SO_2). Even though the smoke and dust debris in the atmosphere was, after 6 months, beginning to settle back to Earth, the SO_2 remained in the atmosphere for another two to three years. The SO_2 now took over the role vacated by the smoke and dust of blocking the sunlight and keeping the Earth in a near freezing state for almost another three years. Finally, the SO_2 began to be washed out of the atmosphere by rain waters. This event itself created a new round of problems for the surviving plants and animals since the SO_2 now fell to the ground as "acid rain" containing huge quantities of sulfuric acid that proceeded to burn the leaves off plants and poison any surviving animals' water holes. Thus, over a period of almost four years following the asteroid impact there was a slow but steady killing of most plants and animals larger than that of an alligator, including the dinosaurs.

Unfortunately, this was again not the end of the story with regards to this extinction event The massive forest fires plus damage to both life and the oxygen atmosphere had also inflicted a major shock to the Earth's normal carbon cycle. Suddenly, the Earth's atmosphere would have contained excessive amounts of carbon dioxide (CO_2) that would now trigger increases in the world's greenhouse effect that would last for possibly another 100 years. Many parts of the world probably suffered from as much as a 20 °C increase in average surface temperatures during much of the next century. If a similar asteroid strike occurred today, our planet would suffer much more than the loss of most of its larger plant and animal species. Earth's entire human civilization, as we know it today, would be virtually destroyed. While millions of citizens all over the world would survive the initial onslaught, their lifestyles would be thrown back into that of the middle ages, or even the stone age. Mankind, however, would persevere and might eventually (after many centuries) be able to return to some semblance of an agricultural/husbandry society, or the beginnings of a new *but quite different* pre-industrial or second industrial revolution stage of development. And, after a much longer period, their daily lives might possibly return to something vaguely resembling the technological civilization they had the day before the "monster from the sky" hit them.

Of course, thanks to the sudden explosion of computers and the space age at the end of the last century, mankind now has the capability of preventing our species from "going the way of the dinosaurs". If detected far enough ahead of time, our scientists now have the technological means of sending unmanned spacecraft to intercept such menacing objects (asteroids or comets) and either physically or gravitationally altering their orbits sufficiently to keep them from striking the Earth.[3]

Advanced technological civilizations on other exoplanets would also likely be able to avoid these types of catastrophes on their own worlds. Still, while an unlikely event, it is possible that large asteroids or comets could, as happened 65 million years ago, again in the future produce another major worldwide loss of life or even a complete and total destruction of the Earth. While virtually all astronomers recognize the very serious nature of this potential danger, very little in the way of preventive actions have been undertaken to make sure this calamity never happens. The U.S. Congress, nevertheless, was apparently aware of this continuing danger from the skies when, in 2005, it issued a mandate to NASA that this agency, by the year 2020, develop a program that would be able to successfully detect and catalogue 90 % of all such threatening objects in our solar system that were large enough (defined as 0.6 miles in diameter or greater) to inflict worldwide destruction and loss of life. NASA now classifies any such objects that are this size or greater as **Near Earth Objects** or "NEOs". Nevertheless, at the writing of this book (2012), no presidential administration has requested, nor has any congress allocated the funds that NASA would need to meet the requirements of this congressional mandate. Since getting hit by a NEO is itself a worldwide concern, it would seem reasonable to make paying for it an international responsibility.

While the primary objective for any NEO protection system will be to provide a means of protecting the Earth from the very worse or global-level impacts, there is still an additional quite serious need for such programs to be expanded to protect us from the much more likely and frequent smaller asteroid or comet impacts that could destroy entire cities. In 1908 a much smaller asteroid, *which is estimated to have been considerably smaller than 0.6 miles in diameter*, exploded in the air over a very remote and unpopulated part of Siberia. Virtually all vegetation and wildlife was destroyed in an area estimated to have been as large as 900 square miles. This asteroid, while not as destructive as the one that killed the dinosaurs, could have destroyed a city the size of modern day Moscow, London, or Washington, D.C. NASA scientists recently estimated that the frequency of asteroid strikes large enough to destroy entire cities ("city-killers") in this size range is approximately one per 100 years.

Close encounters with wandering black holes and rogue exoplanets In addition to being hit by debris (asteroids or comets) left over from the planet's early formation process, astronomers now tell us that wandering asteroids and comets are not the only menacing objects that mankind (or ETs on other planets) need to worry about when it comes to being hit and destroyed by some kind of large object from space (Plait 2008). Astronomers have now identified a potentially "quiet

---

[3] The last thing NASA should try to do is "blow up" a large asteroid or comet using some kind of explosive device. Converting a single large threatening object into a whole truckload (train load) of smaller objects would very likely increase the amount of destruction on the Earth.

invisible" killer that lurks amongst us in the depths of space which the inhabitants of any doomed exoplanet would not even be able to see. When some giant or super-giant stars which are 25 times or more massive than our sun go into a supernova state and explode, they will leave an extremely small but incredibly massive remnant of the star's core behind, i.e., a stellar *black hole*. Because of the extreme mass of black holes, they have a gravitational field that is so strong that nothing, including light, can escape from them. Thus, black holes are always invisible which is why they are so threatening to any inhabited planetary systems they may approach. Many years ago, when first discovered, astronomers estimated that black holes were quite rare in our own and other galaxies. Unfortunately, more recent estimates have increased their potential numbers to a billion or more, just in our own Milky Way galaxy alone. These black holes are definitely moving around out there in interstellar space and could be a threat to any planetary system that they get too close to. Figure 6.4 shows an artist's drawing of a black hole as it might appear while wandering around in a region of space (i.e., a globular star cluster) which contains large numbers of background stars, while Fig. 6.5 shows a view of what could possibly happen when any stars get too close to these "monsters".

Since they are invisible, it would be very difficult for any intelligent living creatures on an exoplanet to detect the presence of a black hole. One possible early sign that a black hole might be headed toward a planetary system might involve a

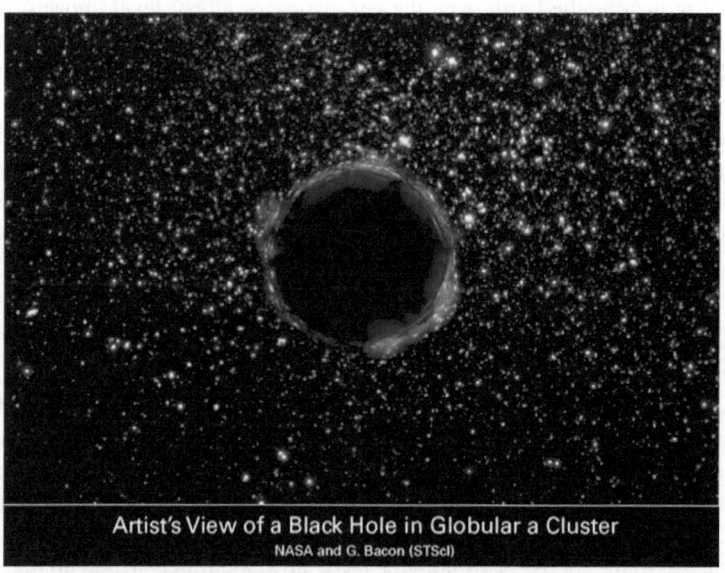

Artist's View of a Black Hole in Globular a Cluster
NASA and G. Bacon (STScI)

Fig. 6.4 Since the gravitational field of a stellar sized black hole is so intense that light itself cannot escape, such objects are invisible. However, if one of these smaller stellar size black holes were located in a star cluster containing large numbers of stars in the background, it might reveal itself by a large circular region which seemed to contain no background stars. Shown is a NASA artist's drawing of what such a black hole might look like (image credit: NASA)

Fig. 6.5 Being the invisible entities they are, the only way astronomers can ever expect to "see" a black hole is by observing how it affects other nearby celestial objects. Shown is an artist's drawing of what happens when an unfortunate star gets too close and begins to have its outer layers sucked up (eaten) by a nearby black hole. As the star gets closer to the black hole, its shape would change from round (*upper left image*) to a *oval* or *oblong shape* (*middle image*) by the gravitational pull of the black hole (*lower right image*) (image credit: NASA)

relatively sudden increase in the numbers of comets that start appearing in the sky and threatening the stellar system's planets. Most planetary systems, like ours, probably contain an extended cloud of dust and debris that extends a huge distance (one or two or more light years) beyond the orbit of the system's most distant planets (e.g., Neptune in the case of our own solar system). These extended clouds, which are called *Oort Clouds* by astronomers, are filled with huge numbers of extremely small icy planet-like objects plus even smaller objects called comets that are slowly orbiting the star(s) of their planetary system. A slowly approaching black hole might gravitationally disturb the orbits of many of these comets or icy mini-planets and fling them in the direction of the system's sun and on a direct collision course with some of the star's exoplanets.

An additional early sign that a stellar black hole might be heading their way might also involve some ET astronomers beginning to notice extremely small changes in the location of other planets or objects (moons, orbiting space stations or satellites, etc.) in their planetary system. The changes would be the result of the black hole's gravitational field beginning to interact with the gravitational fields of these other objects. Depending on how fast or slow the black hole was moving, these subtle gravitational changes might increase slowly or rapidly. Before long the gravitational pull of the black hole would begin interacting with the planet's (and any moons it might have) own gravitational fields to produce larger and larger ocean tides or deadly tsunamis (if the exoplanet, like our Earth, were a water world), and stretching of the planet's interior with resulting catastrophic earthquakes and intense volcanic activity worldwide. Unlike normal earthquakes on Earth that occur infrequently, and in small isolated geographic areas, and which typically last only a few minutes, the black hole generated earthquakes would be virtually continuous, and while generally extremely severe, would vary constantly in intensity. No inhabitants of the planet, except for a few ET astronomers, would have a clue as to what was happening. As the black hole gets closer to the planet, its gravitational field would begin to match that of the planet and things would then really get crazy for any living creatures that might have managed to survive the earlier catastrophic effects. When the strength of the black hole's gravity begins to match that of the equal but opposite pull of the planet's gravity, everything that was not nailed down (e.g., inhabitants, buildings, or transportation vehicles, whatever) would start floating up into the sky along with the rising wind currents of the atmosphere. In addition to virtually continuous severe earthquakes worldwide, the stretching of the entire outer rocky crust of the planet by the constant pull of the black hole's gravitational field would cause the entire planet to begin "popping" like inflated rubber balloons being stuck by sharp objects! Severe and extensive cracks would begin occurring everywhere in the planet's outer surface which would cause the hot molten lava in the mantle, which is normally under high pressure but contained by the solid surface crust of the planet, to begin being released and exploding upwards in the forms of a worldwide frenzy of extremely severe "super volcano" type eruptions that would set the entire planet ablaze. The last stages of this horrific stellar event would involve the black hole beginning to suck in and devour everything that still remained in its path (probably including the planetary system's home star and any other planetary neighbors) and then leisurely continuing on its journey to its next "meal" (breakfast, lunch?) in space.

In addition to wandering black holes, astronomers tell us there is another kind of dead star that could also collide with and destroy planetary systems. Some stars that are slightly larger (4–8 times) than our sun, but not quite massive enough to form black holes will, after they die transform themselves into another kind of nasty stellar remnant. These dying stars are called "neutron stars" by astronomers because when they go into a supernova state, the protons and electrons in their cores are compressed so tightly that they are converted into neutrons. Neutron stars are similar to black holes in being invisible to the naked eye but they may (if they begin spinning and become pulsars) emit radio and X-ray signals which might allow ET astronomers to determine that they are on a collision course with their home planet and, if detected far enough ahead of time, allow enough time for the residents to mount some kind of evacuation plan to prevent the extinction of their species. However, since collisions between stars (whether neutron types or otherwise) anywhere in our vast universe occur no more often than once every 10,000–100,000 years, the reader and the author can probably relax.

Since NASA now has the technical means of altering the orbits of threatening near Earth objects (e.g., asteroids or comets like the one that killed the dinosaurs 65 million years ago), the reader might be wondering whether our scientists and engineers might be able, using unmanned rockets (as they now can in the case of threatening NEOs) shove approaching black holes away from us to keep them from hitting our planet. Today's rocket scientists are totally capable, if they make plans to do so far enough ahead of time, of parking small unmanned rockets close to approaching NEOs and letting the small weight of the rocket serve as a counter-weight to gravitationally alter the NEOs orbit just enough to cause it to miss us. Unfortunately, even though black holes are typically small (being no larger than an asteroid) they are "extremely" heavy with many being at least as heavy (massive) as whole stars. It would, therefore, be totally impossible to build a rocket that would be large enough to have any effect on the gravitational field of an approaching black hole or even a neutron star. Of course, advanced intelligent ETs on other worlds might be able, using some kind of powerful rocketry or laser beam technologies (or wormhole technologies?) to successfully shove threatening black holes or neutron stars out of the way. However, the only thing the early detection of a menacing black hole by Earth's present astronomers might allow would be for our scientists and engineers to hurriedly build large Ark-like space ships, load them with lots of supplies plus a few (very happy) winners of an "Adam and Eve" lottery drawing and making a mad dash for the closest star that hosts an exoplanet that is similar to our own now doomed home planet.

However, even if the wandering black hole or neutron star does not directly strike a stellar system and destroy all or most of its planets as well as its home star, a "near miss" scenario would itself, because of the intense gravitational disturbance that these threatening stellar remnants produce on the orbiting planets of the stellar system, disturb the normal orbital relationships among the different planets and cause one or more of them to either collide with each other or, as we will see next, cause them to be tossed out of their stellar system to become what astronomers call "rogue wandering planets" that could then pose threats to other stellar systems in their galaxy.

Whether caused by close encounters with wandering black holes or other kinds of disturbances, astronomers now believe that the possibility that planets could be ejected or tossed out of their own solar systems and subsequently wander across space and crash into another solar system is a much greater threat than previously considered possible. In 2012, researchers from the Kavli Institute for Particle Astrophysics and Cosmology at Stanford University made the bold prediction that our Milky Way galaxy may contain as many as 100,000 wandering exoplanets that were previously ejected from their home systems. Other astronomers have subsequently suggested that such rogue wandering exoplanets may actually number in the millions or billions in our own galaxy (Fig. 6.6a, b). One reason that scientists now believe rogue wandering planets may be more common than previously thought is that the so-called "hot" super-Jupiter exoplanets (i.e., large gas giant planets that form in the colder distant regions of their stellar systems but later spiral in towards their home stars) that the new planet hunters are finding are believed to be capable of gravitationally flinging some of their system's inner rocky planets out of the stellar system when they move closer to their home star. If they do exist, one of these wandering nomads might someday enter our solar system (or another stellar system) and collide with one of the system's own resident planets or, as we will see next, create some kind of "gravitational nightmare" scenario.

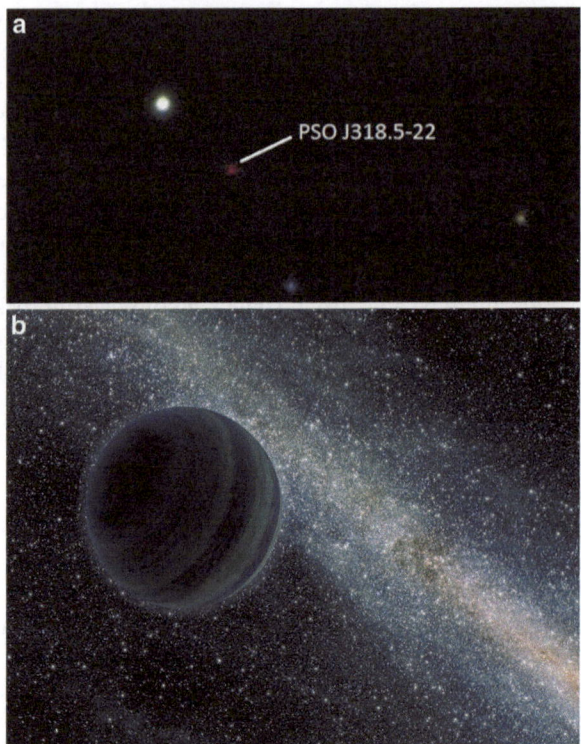

Fig. 6.6 In recent years, astronomers have suddenly started discovering that exoplanets rejected from other solar systems that are forced to wander about our galaxy without any home stars to orbit around is a far more common phenomenon than previously believed. (**a**) Shows a photograph of one such lonely ("orphan") Jupiter sized exoplanet that was recently discovered by Canadian astronomers (image credit: N. Metcalfe & the Pan-STARRS 1 Science Consortium). (**b**) Shows an artist's drawing of what a Jupiter sized roaming exoplanet might look like close up. Although all ten of the rogue exoplanets that astronomers have discovered so far are Jupiter size or larger, it is likely that astronomers will begin finding smaller exoplanets as our search techniques improve (image credits: Wikipedia and NASA/JPL)

All planetary systems are threatened whenever the relationships between the gravitational fields of their planets are severely disturbed A major reason our solar system or any solar system is so vulnerable to the sudden arrival from space of huge foreign objects is that all stellar systems that host multiple circling planets are constantly, throughout their entire lives, engaged in an incredibly delicate and fragile "balancing act" with respect to how the orbits of individual planets constantly interact with and affect the orbits of all of the other planets in their stellar systems. In all planetary systems, planets with different masses are constantly orbiting their home star(s) at different distances and at different speeds. The gravitational field of any planet directly affects the exact manner in which any other planet executes its orbit. A slight alteration of the orbit of any particular planet can and does directly affect the orbits of all of the other planets. If anything happens to alter, even briefly, the gravitational field of a planet, it may create a "ripple effect" in the orbits of the other planets, sometimes with dire consequences. If external objects, such as the small dense cores (known as white dwarfs) that are left behind following the death of stars similar in size to our sun, or rogue exoplanets that are ejected from other stellar systems were to enter our solar system and either slam into or get too close to one of our planets (such as our resident "body guard" for such calamities, i.e., Jupiter), it might cause the orbits of the other planets to be altered enough to trigger some of them into being tossed out of the solar system or to collide with one of their planetary neighbors. Even if the rogue exoplanet (or other now dark "white dwarf" stellar remnants) do not directly collide with any planets in the course of their invasion of another planetary system, their huge gravitational fields could create havoc with the gravitational interactions among these planets and still possibly cause some of them to be tossed out of their stellar system. Astronomers, therefore, now believe that many planetary systems in the universe remain in a threatening "shooting gallery" type situation long after they are somewhat relieved from the worst threats of being hit by leftover debris that occurs during their early formation or heavy bombardment periods. NASA may need to take steps to protect us not only from threatening asteroids or comets in our own stellar systems but perhaps also threatening "junk" delivered courtesy of other neighboring planetary systems (or dead stars).

In addition to the possibility of some planets being completely tossed out of their solar systems as a result of an "invasion" by external foreign objects, some astronomers believe that another type of hazard to the stability of planetary orbits may involve occasional erratic alterations in the orbits of some planets relative to other planets. Astronomers note that the present day orbits of Neptune and Uranus are different from what would be expected on the basis of the subtle differences in size, eccentricities, and distance of these two planets from the sun. Both planets are slightly more massive than would be expected based on the normal decrease in the amount of building materials that would be expected to be available to build planets this far out in a planetary accretion disk. Some scientists, therefore, believe Neptune and Uranus originally formed closer to the sun, as did also Saturn, and that the gravitational field of Saturn (along with that of Jupiter) may have, in combination with disturbances from passing comets, slightly altered the orbits of these two planets causing both of them to move in the direction of the outer solar system. As these planets moved to the outer regions of the solar system they may have become slightly more massive as a result of picking up additional building materials from other smaller left over protoplanets as well as objects in what is now the asteroid belt. In addition to being shoved further out from the sun, it is also possible that the relative orbits (i.e., distances from the sun) of Neptune and Uranus may have also been reversed on several occasions in the past.

Some astronomers, however, now believe that this inherent instability of the orbits of the planets in our solar system might have been much worse about four billion years ago just following the completion of the accretion of our present cadre of eight planets. At that time the orbits of the planets were more unstable than they are today. In addition to being a major threat to the long term survival of our solar system, this situation may have had a huge positive influence on the development of life on Earth. At this earlier time, the orbits of our two largest planets, Saturn and Jupiter, were closer together then than they are today. This resulted in these two giant planets periodically coming close to each other while orbiting the sun and producing a much stronger combined gravitational field. This unusually strong gravitational field not only caused Neptune and Uranus to be pushed further out in the solar system but may have also created havoc for the orbits of any nearby objects in the asteroid belt causing huge numbers of them to break out of their normal orbits and collide with the Earth and other members of the rocky inner planet group. Following their initial accretion, all of the inner planets were still extremely hot, molten, and very dry. Since many asteroids (then and now) are known to contain large quantities of water, these wandering asteroids may have been the source of all the water that now fills our oceans (and also similar oceans on Venus many eons ago before the sun warmed up and vaporized them).

Deadly radiation from nearby or distant exploding stars   In addition to threats from external physical objects, another very ominous threat (because it is invisible and currently totally undetectable prior to its occurrence) are *gamma ray bursts* (GRBs) that are commonly associated with the traumatic explosions of some of the most massive known stars.[4] Although stars the size of our sun do not explode at the end of their lives, giant and supergiant stars end their days in catastrophic explosions. The explosions of the giant stars are less intense and are called **supernovas** by astronomers, while the death of some of the largest of the supergiant stars are extremely violent and are labeled as **hypernovas**. In fact, most astronomers believe that the total energies involved in many hypernova explosions are second only to the Big Bang event itself. Most astronomers now believe that GRBs are the most intense stellar explosions that can occur in the universe and most are located far beyond our own galaxy (millions of light years away). GRBs are believed to be incredibly rare phenomena with no more than one or two occurrences per day somewhere in the universe. Since most of the energy from GRBs is gamma radiation and virtually none is visible light, some other astronomers now believe the total number of GRB events may be closer to 100 per day, but because of their incredible distances and being invisible to our telescopes, they are not seen. Some of these distant hypernova stars are 600 times or more massive than the giant stars that elicit supernovas. When stars of this size explode, i.e., "go hypernova", their central cores shrink to a diameter of 2 or 3 miles or even less, while the mass of the core remains the same as the original star. This small extremely compact remnant of the original star may then become one of those dreaded objects that astronomers

---

[4] Astronomers classify stars as belonging to three basic size groups. The smallest stars (those the size of our sun or smaller) make up about 90 % of all stars, while the "giant" stars with diameters from 10 to 100 times greater than our sun make up about 9 %. Finally, the rarest and the biggest stars, called the "supergiant" stars with diameters that can be as much as 600 times or greater than our sun make up less than 1 % of all stars.

refer to as invisible stellar ***black holes***. Unlike smaller supernova explosions which produce shock waves that completely blow off the outer parts of the star, the explosions of many supermassive stars do not blow off the star's outer layers. The material of the outer layers falls or spirals into the black hole and its gravitational energy becomes converted into heat and intense radiation. Incredibly intense but relatively narrow beams of gamma rays are shot straight out into space at close to the speed of light from both the north and south poles of the black hole (Fig. 6.7). If an exoplanet of any stellar system is located directly in the path of either of the two GRB beams and, at a relatively close distance (anything less than 5,000–6,000 light years is considered by many astronomers as probably "dangerously close"), it will be hit by the beam with devastating consequences. If the hypernova explosion is relatively close (i.e., less than just a few light years away), the exoplanet would immediately become "toast" and be totally vaporized! If it is several thousand light years away, the inhabitants (the seeing and knowing ones) of the exoplanet would not be able to see it since most of the electromagnetic energy associated with GRBs are in the gamma ray range which is invisible to human eyes. Since most of the gamma rays would not reach the ground, the major effect of the GRB burst would be a rapid and total destruction of any oxygen ozone layer the planet might possess which would cause all exposed life forms on the planet to immediately begin being bathed by intense ultraviolet light, with all the dire consequences that would entail, including sunburn, cancer, radiation illnesses, and eventually death. Unlike a normal supernova event that produces lots of visible light, which the residents of the planet might interpret as the sudden appearance of a second (or third) sun in the

Fig. 6.7 Artist's drawing depicting a gamma ray burst event associated with the hypernova explosion of a dying supergiant star. The exploding star emits focused beams of deadly gamma radiation from both its north and south poles that travel at speeds close to that of light. If an exoplanet of any stellar system lies in the path of one of these two bursts, and at a distance of 6,000 light years or less, the atmosphere and all exposed surface life would be immediately destroyed (image credit: NASA)

sky, the GRB source would be invisible. It does seem, therefore, that GRBs are nature's ultimate "silent doomsday machine", if they hit you! However, being the incredibly rare and powerful but highly focused monsters they are, science's "deep time" concept once again keeps most of us from getting overly stressed out at the thoughts of such a catastrophe!

GRBs were discovered accidentally during the Cold War with Russia when the U.S. military launched special satellites into Earth orbit to monitor for the release of gamma rays produced by Soviet nuclear bomb testing. Following the signing, in 1963, of the *Nuclear Test Ban Treaty,* the U.S. military was concerned that the Russians might attempt to secretly continue their nuclear weapons testing program by launching such devices with rockets and detonating them in space. The U.S. satellites did, in fact, begin detecting the presence of gamma rays originating from various locations in the sky. Fortunately, our astronomers were able to quickly confirm that they were the product of distant hypernova events rather than Russian testing.

It is interesting that scientists at NASA and the University of Kansas, led by A. Melott and B. Lieberman, have recently proposed the idea that a GRB may have actually hit the Earth about 440 million years ago between the Ordovician and Silurian Periods of the Paleozoic Era, causing the first of our planet's five major mass extinction events (Melott and Lieberman 2013). A supergiant star 15 times or more massive than the sun, located in a region of our galaxy possibly as far away as 6,000 light years, may have exploded sending a GRB our way that hit us at that time. The passage of the GRB would have lasted no more than 10 s but it could have resulted in a near total destruction of our protective ozone layer that lasted for at least five years. Since most life at that time was confined to the oceans, only surface or near-surface ocean life (e.g., planktons) would have been affected. This, however, may have produced a ripple effect in the food chain that resulted in the destruction of at least 70 % of the deeper and larger forms of sea life as well. A severe ice age phenomenon is also believed to have been associated with this extinction event. In addition to destroying the ozone layer, the GRB beam could have caused the creation of huge quantities of nitrogen dioxide, a gas similar to the smog we frequently see in today's polluted cities, which would have blocked the sun and produced near freezing temperatures.

The dreaded hypernova explosions associated with the largest of the supergiant stars, therefore, are so powerful that they have the potential of possibly destroying life on exoplanets located in far distant parts of their own galaxy or even beyond their galaxy. The supernova explosions associated with the smaller giant stars can, however, also produce huge amounts of deadly X-rays and other forms of ionizing radiation (e.g., ultraviolet and gamma ray) that may create havoc for any nearby extraterrestrial life located on exoplanets in the same galaxy. In contrast to the extreme rarity of hypernovas, the number of occurrences of supernova explosions is about one every 50 years in our own Milky Way galaxy. Figure 6.8 shows a photograph taken by astronomers in 1987 of a supernova explosion that occurred in the Large Magellan Cloud, which is a dwarf galaxy that is only 168,000 light

Fig. 6.8 In addition to gamma ray bursts associated with the most massive dying supergiant stars, many smaller giant stars will also end their life spans with supernova explosions. These explosions can also emit radiation in all directions that can be deadly for any lifeforms on nearby exoplanets. It is extremely rare to be able to obtain a photograph of a star that is still in the early stages of a supernova or hypernova explosion (astronomers typically can only photograph the remnants of such explosions hundreds or thousands of years following their explosion). This photograph, taken in 1987, shows a supernova event as it occurred in the Large Megellan Cloud, The Magellan cloud is a dwarf galaxy that is located 168,000 light years from our Milky Way galaxy (image credit: David Martin, Angelo-Australian Observatory)

years distant from the Milky Way galaxy. These exploding stars, unlike their larger GRB eliciting cousins, may announce their occurrence by the sudden appearance of a new bright star in the sky, which might even be bright enough to be seen during daylight hours. Instead of the extremely intense and narrow GRB beams emitted by many larger hypernovas, these smaller supernova events will also emit diffuse patterns of gamma radiation plus other kinds of deadly ionizing radiation (e.g., X-rays) in all directions. Typically, these smaller supernova explosions would have to be located much closer to a planet to be lethal for any living inhabitants.

While nearby exploding stars (either super- or hypernovas) would be bad news for any exposed inhabitants of any nearby exoplanets, things could be worse depending on where the exoplanet is located within its home galaxy. If the planetary system was located more towards the center of its galaxy, its immediate neighborhood would be much more crowded by other stars. More stars would mean a greater likelihood of the occurrence of nearby supernova events. Fortunately, our own home planet is located out in the "boom-docks" or suburbs of our Milky Way galaxy where we have far fewer exploding stellar neighbors to worry about.

Smaller stars like our sun that do not explode as supernovas may also threaten inhabitants All stars are born, grow old, and eventually die. Unlike the large (and rare) giant and supergiant stars which end their days via sudden massive explosions, smaller stars like our sun (which constitute approximately 90 % of all stars), will end their lives, not in a sudden explosion, but in a slower and more gentle expansion and expulsion of all their matter into space (Spangenburg and Moser 2003; Lequeux 2013). When such smaller stars begin to run out of the hydrogen fuel they need to continue shining and start to die, they will begin expanding in size and initially become a red giant type star and eventually grow into a huge gas cloud that scientists refer to as a ***planetary nebula*** (Fig. 6.9). The size of the red giant star will get so great that its outermost surface will eventually engulf any closer planets (like the Earth) that orbit it and burn them up (Fig. 6.10). As the outer layers of the star begin to expand, the core of the star will collapse into a very small and extremely dense object that astronomers refer to as a **white dwarf**. The white dwarf remnant left behind from the star's slow expansion will continue to shine for a few million years but since it is now far too small to allow nuclear fusion, it will eventually become a small extremely dense dark object. Since the death of sun-like stars is very slow, and always involves a gradual heating up of the star as the process continues, things will probably get so hot on any inhabited planets that all life will perish long before the dying process is completed. Thus, any ETs still living on any planets circling such sun-like stars, will face this inevitable fate, unless they (or us) manage to survive long enough and get smart enough to relocate their species to other friendlier exoplanets in nearby stellar systems or now warmer moons (e.g., Europa or Titan) located further out in our own solar system.[5]

However, smaller stars like the Earth's sun do not have to wait until their final death throes to also pose a significant threat to the well-being of any intelligent life forms that might still inhabit them. Any advanced civilization that has managed to reach the stage of being totally dependent on electricity for everything they do might also wake up one morning and find that their friendly warm sun has suddenly turned into an unfriendly monster of some kind. Stars similar in size to the sun periodically exhibit unique physical features on their surfaces that astronomers call "sun spots" (Fig. 6.11). These spots, which exhibit sizes that can be larger or smaller than the diameter of the Earth, can be numerous or small in number, and are always darker than the surrounding surface of the sun. The reason they appear darker is because the spots are always cooler than the adjacent sun surface. While the surface of the sun might be 9,000 °F, the sun spots might be thousands of

---

[5] Unless advanced multi-cellular creatures like us manage to find a way to escape from our hot sun and move to cooler friendlier homes located further out in our own solar system or to exoplanets circling other stars, life on our planet will eventually succumb to the excessive heat from our sun and all life will perish on our planet. It is interesting that life on our planet first arose as single-cell organisms (who liked it "hot") circa 4 billion years ago, and the last life to survive on our hot planet will also be single-cell organisms who, once again, will "like it hot". Because of the evolutionary heating of our sun, life on our planet will slowly revert back its former single-cell thermophile format before finally dying all together.

Fig. 6.9 Stars close to the same size as our sun or smaller do not end their days by horrific supernova explosions but by a more gentle expansion and expulsion of all their matter into the surrounding interstellar space except for a small dense core known as a white dwarf. These dying stars are known as "planetary nebulae" since they resemble planets more than some kind of explosion event. Shown are four typical planetary nebula. The small white object at the center of each image is the white dwarf remnant that is left over following the expansion of the surrounding nebular clouds from the dying star (image credit: NASA, ESA, and the Hubble Heritage Space Telescope Team)

Fig. 6.10 In a few billion years, as our own sun starts to die and begins expanding into a red giant star, it will get so large that its outer surface will get very close to Earth or perhaps even engulf it and burn it up (image credit: Wikipedia Commons)

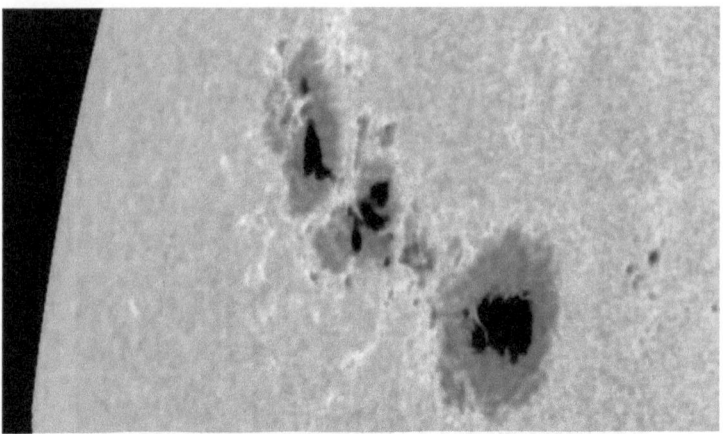

Fig. 6.11 Stars similar to our own sun go through cycles in which they periodically display darker regions called "*sun spots*" on their surfaces which are areas of intense magnetic solar storms that, while not a major threat to life, can be real bad news for the inhabitants of any exoplanet that may be totally dependent on electricity to keep their complex technological based civilization running. © NASA/SDO/HMI (image credit: NASA)

degrees cooler. These sun spots are the centers of ongoing "solar storms" involving intense magnetic activity which results in the ejection of large clouds of solar matter referred to as "solar flares" or "prominences" (Fig. 6.12) that contain incredible numbers of electrically charged atomic particles that can drastically interfere with any electrical or communication systems that the inhabitants of any exoplanet may have developed. These flares can "fry" the internal electrical circuits of communications satellites and all other telephone, internet, and any mechanical systems or power stations worldwide that require electricity to work. Although, for our sun, sun spot activity is known to vary in intensity over approximate 11 year cycles (plus other even longer cycles), there is no way (yet) to know in advance when the activity will get intense enough to drastically disrupt the ongoing lifestyles of the planet's inhabitants. A really severe bout of sun spot activity occurred on 1 September of 1859 that interrupted electrical telegraph communications and produced intense and colorful aurorae ("northern light shows") over both the northern and southern hemispheres. Except for the colorful light show in the sky, this event went virtually unnoticed by the world's population since mankind was still far more dependent on oil lamps and wood stoves than on those few telegraph wires that were strung around some cow pastures. A similar sun spot outburst would today, because of mankind's total dependence on electricity for virtually everything we do, create a major worldwide communications and technological meltdown that would take many years from which to recover.

Unfortunately, in addition to creating havoc with any ETs electrical systems, solar flares also have the potential of destroying their planet's atmosphere. Fortunately, for any residents of rocky exoplanets that are close in size to or larger than our planet Earth, nature has provided an excellent "protective shield" that causes

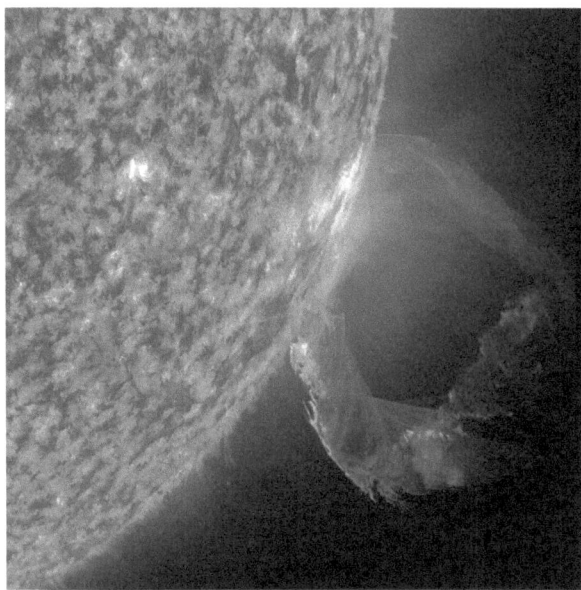

Fig. 6.12 In addition to sun spots, stars also frequently show disturbances involving intense clouds of atomic particles called *prominences* being ejected from their surfaces and thrown out into space. If these clouds of electrically charged particles (or ions) reach the Earth they will produce intense and colorful displays in the upper atmosphere called auroras ("northern lights") over the northern atmosphere and also sometimes close to the south pole. These charged clouds can cause havoc with the electrical circuits of orbiting satellites as well as electrical power stations or electrical grids on the planet's surface (image credit: NASA)

most of the deadly ionizing radiation (e.g., cosmic rays, UV light) from any solar mega-flares to be harmlessly deviated around the planets instead of entering and, over time, stripping away the upper regions (e.g., ozone layer) of their atmospheres. Earth today still has a small solid iron core that is surrounded by a larger liquid iron core. When our planet rotates on its axis the liquid outer core tends to lag behind the faster rotating solid inner core to produce an electromagnetic force field which prevents any ionizing radiation that is released from the sun from attacking our atmosphere. Planets that are smaller than our Earth probably also possess molten cores in the early years following their formation but typically lose these cores faster than larger planets because of their smaller mass and gravitational fields. The planet Mars is a good example of this. Mars may not today have a molten core that can produce a protective magnetic shield and, as a result, has had most of its atmosphere stripped away over time by solar flares. In its youth, three or four billion years ago, Mars was possibly still hot enough to have a molten core that could have allowed it to briefly support a thicker and more life friendly atmosphere. Mars today has lost much of its atmosphere. The air pressure at the surface today is about 1/100th that of Earth's sea level pressure.

Having described how Mars probably lost much of its atmosphere as a result of its small size, I must take a brief detour from the topic of this chapter related to how life on any planet may be vulnerable to external factors (forces) and briefly describe what our life scientists believe are probably other major physical characteristics of planets that may pose threats to the development of any life forms that might be similar to those that our scientists are familiar with. In addition to *size*, other critical features of exoplanets, including their *physical composition* (e.g., whether smaller rocky or water planets, or large gaseous planets) and *locations within their own stellar systems* (e.g., whether they orbit in their home star's temperate habitable or "goldilocks" zone which would allow the presence of liquid water on their surfaces), are also believed to be critical for the development and survival of any viable life forms that our life scientists presently think are possible. In addition to *size, location,* and *composition* one additional critical feature of planets that is likely to be critical for determining whether they might be life friendly has been identified by the planet hunters and that is *history* (e.g., whether a super-Jupiter or even smaller planet formed further out in its stellar system and then subsequently migrated to a hotter close-in orbit, or vice versa) might also determine whether or not any particular planet (or its moon) would have enough time to allow the incredibly slow (long term) evolutionary changes that are critical for the development of life.[6]

Thus, planetary size appears to be only one of many critical factors that determine whether life may or may not be able to evolve on exoplanets. It may, however, be especially important for determining whether dense enough atmospheres (whether dominated by oxygen, carbon dioxide, or other kinds of gases that might conceivably be needed by other kinds of presently unknown life forms) can occur that our life scientists currently believe is critical to allow at least carbon-based and water dependent life to evolve. The larger size of our own home planet was probably an important factor in determining why life on Earth was able to evolve to an advanced multicellular level, and why life on Mars, if it developed at all, may not have been able to advance beyond an early single-cell microbial stage due to the premature demise of its atmosphere.

The Kepler space telescope astronomers now tell us that large numbers of their newly discovered exoplanets are significantly larger than our Earth (i.e., with diameters as much as five or more times that of Earth). Many of these more massive planets, however, are rocky planets and not gas giant planets. Because of being both rocky plus massive, these planets would have greater gravitational fields and perhaps more dense atmospheres. A thicker atmosphere might possibly cause them to have a greater greenhouse effect which would require them to orbit at a

---

[6] Another way in which "history" can affect the longterm habitability of planets relates to the possible universal tendency of stars to very slowly heat up as they age. In its younger days, i.e., 3 or 4 billion years ago, Venus may have been a life friendly water world. With the subsequent slow warming of our sun, Venus became too hot (all water evaporated, and what may have been a life friendly atmosphere turned into an extremely hot and deadly "gas chamber" containing a very hot and thick cloud covered atmosphere containing nasty sulfuric acid and carbon dioxide/nitrogen gases).

greater distance from their home stars in order to maintain cooler temperate climates which allow liquid water to exist on their surfaces. Some of these more massive rocky planets might also have more energetic plate tectonics which could allow more efficient recycling and replenishing of the elements and chemicals (such as carbon and nitrogen that are critical to life; at least as we know it on Earth). This might also allow some of these larger rocky planets to host more advanced multicellular forms of life (including intelligent life) for a much longer period of time than would be possible for smaller planets like the Earth. This "longer life friendly" scenario, however, would probably work better if the larger exoplanet's home star was a long lived and stable red dwarf and not a short-lived unstable giant star (see reference by Stevenson 2013, re. "life around Crimson suns"). Of course, any living surface creatures of such large planets would probably need strong limbs to hold them up and/or larger wings to assist them in flying around in a thicker and possibly more stormy and turbulent atmosphere.

Thus, in just four short years of active exoplanet searching, the one thing the Kepler space telescope people have done for us is to begin demonstrating that, while exoplanets are very common in our vast universe, those that, at their present stage of evolutionary development, may be close to being twins of Earth may be more rare then some of us expected (see references by Jayawardhana 2011 and Mason 2008).[7] For those of us who are diehard "trekkies", i.e., serious fans from the Star Trek television generation, these preliminary new findings may come as a bit of a disappointment. If twin Earths are extremely rare, then intelligent ETs that are, in any way, similar enough to us with whom we might be able to easily communicate and socialize with might also be few and far between. On the other hand, primitive single-cell microbial life forms, similar to Earth's extremophiles, who can evolve more easily and quickly and seem tougher and more capable of living in far more hostile environments then we wimpy earthlings might be *everywhere*, including many of those huge, cold, hot, and freakish exoplanets that our astrobiologists are now telling us may also be *everywhere*. Perhaps the biggest task our future astrobiologists will face, given this amazing variety of exoplanets, may not be

---

[7] One thing that man's incredibly short biological lifespan has done for most of us, and especially for the non-scientists among us, is to make us insensitive to the very real fact that virtually all of the larger physical systems elsewhere in the universe (planets, stars, galaxies, etc.) have unbelievably long "life spans" and go through unbelievably slow changes (ergo, the astronomer's use of the frustrating concept of "deep time").On Earth, for example, man was preceded by tiny heat loving microbes billions of years ago and, because the sun is heating up, we will again be replaced by such single-cell microbes before our sun meets its final demise in another 4 or 5 billion years. Life changes as its physical environment changes, and the moons of some of our own outer planets may eventually develop multicellular carbon-based life forms as our sun heats up and converts their frozen water into rivers or oceans. Europa and Titan may undergo this change long after man has either burned up or managed to escape from our dying planet. When the Kepler planet hunters discover what now appears to be exoplanets that are probably totally hostile to our form of life, they need to keep in mind that this situation might have been entirely different many eons earlier or may be totally different in the far distant future when the exoplanet evolves into a place that would be more capable of hosting more advanced life forms of some kind.

just finding life but to actually reevaluate what life is, or is not, as well as how dramatic differences in planetary physical/chemical histories effect whether life forms can or cannot develop, plus exactly what kinds of life can or cannot develop or evolve (Chela-Flores 2001; Gargaud et al. 2009; Holley 2012; Irwin and Schulze-Manuch 2010; Schultze-Manuch 2013; Toomey 2013). The Kepler scientists are now telling us that the differences between living and non-living matter in the universe is probably *not simple*! Man's long term belief that there are only two kinds of physical matter in the universe, i.e., living or non-living, may now be headed to history's "recycle bin" alongside the idea that we are alone in the universe.

Finally, having made this brief "detour" into the topic of how the basic physical characteristics of planets are undoubtedly so important to determining whether life may or may not be able to develop, the author will now briefly return to the important topic of the present chapter related to the discussion of possible external threats to life on our planet as well as other exoplanets, Although frequently considered to belong more in the realm of science fiction than real science, the author believes he must briefly discuss one last external threat to life in the universe that, up until the last few years, was ignored by most scientists but did allow our science fiction cohorts to make lots of money. Since growing numbers of scientists now believe that ETs may be out there in space, some are now beginning to consider the possibility that some ETs may themselves be hostile to our form of life.

Potential threats from unfriendly extraterrestrial visitors Hollywood movie makers and science fiction writers have, without a doubt, had much to say about this topic. Could mankind wake up some morning and find itself engaged in some kind of *war of the worlds*? This is a topic that no scientist on Earth knows anything about, but every science fiction writer believes could happen. Among those scientists who believe intelligent life forms probably do exist on other exoplanets, some welcome the thoughts of close encounters with alien life forms, while others totally fear it. Will they, as the old Hollywood cliche goes, *"come in peace or to destroy"*? Many eminent and well respected scientists, including Steven Hawking, believe they very well may come to destroy, pillage, or steal something from us, rather than coming to save us from our "depravities". Since mankind is the only intelligent life form the author knows about, I unfortunately lean towards Dr. Hawking's position. Our species evolved directly from predatory ancestors whose day-to-day survival depended on their ability to kill or be killed. Although our species has managed to give birth to many remarkable non-predatory citizens such as Mother Teresa, Mahatma Ghandi, Martin Luther King, Mark Twain, Nelson Mandela, etc., our species' track record in the area of fostering "kind and good people who only want the best for their fellow humans" is far from stellar. From ancient times to the present, every few years our citizens go to war and try to kill each other, and the single most expensive item in the national economies of many countries on Earth continues to be the building and stockpiling of weapons of mass destruction with which to kill our fellow humans. Therefore, mankind definitely has a long history of _not_ coming in peace whenever his own tribe first encounters members of other tribes living on the other side of the mountain, or on the other side of the ocean.

Since our scientists know nothing about where in the universe any possible alien visitors and/or invaders may live, or how smart they may be, or whether they are living creatures or machines, or what kinds of space ships they may travel in, or why they might either want to be our friends or our destroyers, we would most definitely be totally unable to speculate as to their possible reasons for visiting, invading, or destroying us. And, considering the fact that the only intelligent civilization we know of, i.e., **US**, is still possibly a newborn infant in the universe of advanced intelligent life forms, we would not know, even if ET did manage to survive their own unruly adolescent years, whether they might consider us intelligent enough to even bother with, except possibly to swat or stomp on like we do with our own insect relatives. As for possibly invading our planet to steal our gold, uranium, diamonds, or other rare and valuable commodities, why would a life form that is intelligent enough to have avoided their own self-destruction in their earlier years, or is now smart enough to build space ships capable of cruising back and forth halfway across the universe, not have the technologies they need to create, build, or acquire anything and everything they might need or want to make their lives more pleasant and comfortable.

And, of course, the idea that ET might come here and destroy us to prevent us from threatening their own home planet with our nuclear weapons is also (maybe) the product of overly imaginative science fiction writers, although it does sell books and movie tickets. Thus, it is highly unlikely that biological life forms that managed to survive their own predatory early years might land on the White House lawn and hold a press conference to tell us that if we do not rid our world of nuclear weapons, they will rid us of our world. Still, the author believes that one possible reason for aliens "invading" us might be that our planet and its environment may be virtually identical to that of their home planet which they have been forced to abandon because of the threat of a dying sun or some other type of imminent catastrophe. If such ETs are carbon and water-based life forms very similar to us, they might find it more economical to search for twin-Earth types of exoplanets that they could move into that would require minimal alterations (re-modeling). Rather than choosing a different kind of exoplanet with a different environment which would require extensive, expensive, and time-consuming terra-forming (i.e. completely re-tooling, re-engineering, or altering the planet's environment), or even using their own advanced medical technologies to re-engineer and alter the biological makeup (genetics, biochemistries, etc.) of their species, it might be easier just to "move in". And, if the ETs have developed a technology that makes rapid and efficient space travel possible (perhaps at close to light speeds or beyond), they might build huge Ark-like vehicles (or gigantic flying "mechanical" cities)[8] that

---

[8] A few scientists have suggested that asteroids which have now been found to also exist in other stellar systems, might make an excellent "vessel" for any inhabitants that need to escape their dying planetary system. Our solar system, as well as many others, contains a huge number of

could keep their citizens comfortable and happy for however many generations it might take to locate such twin Earths (or maintain them in some form of long term suspended animation). Hopefully, such ETs might be kind enough not to destroy and replace us but choose to negotiate some kind of co-habitation arrangement in exchange for a "rental fee" involving sharing their advanced technologies with us to make life even better for everyone. Unfortunately, many scientists, including the author, believe it is more likely that any such advanced aliens might be similar to us in sharing a strong predatory history which could motivate them to simply exterminate us.

In my recent book (Cranford 2011), I presented a relatively short summary of my own personal review of the published evidence for the existence of possible extraterrestrial visitors from space, or what is commonly referred to as unidentified flying objects, UFOs, or "flying saucers". On June 24, 1947, a private pilot named Kenneth Arnold observed a group of nine strange "silver disks" flying in formation at supersonic speeds (estimated at 1,200 miles/h) close to his airplane near Mt. Ranier in Washington state. Later that day, Arnold told a newspaper reporter that he had seen several strange objects flying at high speed which resembled "dinner plates or saucers that someone was skipping over the surface of a lake." The reporter immediately wrote a story about the sighting and submitted it to his news service. The next day, newspapers all over the country ran a news article describing how strange flying saucer shaped objects had been seen in Washington state the previous day by a private airplane pilot. This flying saucer story immediately launched what can only be described as the beginning of a worldwide flurry of sightings of strange flying objects ("flying saucers") that has continued to the present time. While some people believe these objects are weather illusions (strange cloud formations or lightening), flying birds, or normal commercial or military aircraft, or other normal celestial objects (stars, planets, meteorites, comets, etc.) that they cannot identify, many others have classified them as space craft being piloted by living creatures from other worlds in the universe The author suggests that the reader look at the two books by Bennett (2008), and Kean (2010), referenced at the end of this book for two excellent summaries of the pro and con "controversy" that sightings of UFOs has triggered among the world's population. While I have never, even as a lifelong amateur astronomer, witnessed any aerial

---

relatively "small" asteroids in their asteroid belts that are constructed of rock and metal that could be converted into excellent "escape vessels". In addition to being made of rock and metal, many of these asteroids contain large quantities of water which would also come in handy for any future escapees. If the citizens of a doomed planet were to become aware of their need to escape and had sufficient time to prepare, they could send hordes of engineers and workers (probably robotic types) into their asteroid belt to hollow out the interiors of a large number of the larger asteroids to convert them into homes for the escapees. They could then link or chain a large number of these converted vessels together and attach gigantic solar sails or some kind of nuclear rocket engines and launch these huge "mechanized cities" toward the nearest life friendly stellar system. Perhaps, in the distant future, the historical "rallying call" that American settlers or pioneers used in the early nineteenth century when relocating to California by wagon trains of *"Westward ho, the wagons"* will become *"Starward ho, the asteroids"*!

phenomena that I would personally classify as a probable or even possible space craft from another world, as a scientist I believe that the amount of evidence that has been gathered over the years by reasonably reliable witnesses makes the possibility of such extraterrestrial visitors very real. My personal opinion, however, is that these UFOs, if their origin is indeed from other exoplanets, are not being piloted by living (biological) creatures but instead by some kind of remote controlled or independently functioning "mechanical" devices (auto-pilots, robots, or other forms of advanced artificial intelligence). These space craft, therefore, are more likely space probes carrying forms of automatic remote sensing technology (e.g. cameras) that are gathering information about our world and sending it back to their home planets. The fact that there have been reports from pilots of military and commercial aircraft that these UFOs do not appear to be hostile in any way but can "interact" with our airplanes to avoid mid-air collisions suggest that they are functioning independently and are not totally under the control of distant control centers (exoplanets, or even "mother ships" of some kind). While this may suggest that the builders of these machines are not predators (or have chosen to trade in the predatory thing for a more rewarding and lucrative high technology lifestyle) out there looking for new worlds to destroy or conquer, it could be that they are searching and have concluded (or will conclude) that our planet is not worth invading or cannibalizing. Then again, our planet and its occupants may be the target of a very lucrative ET tourist industry that is entertaining their citizens with the antics of a very entertaining bunch of bizarre life-fellows. The "Amos and Andy", and "I love Lucy" radio and TV shows we humans inadvertently beamed out into space many years ago may have attracted the attention of the bosses of some ET Entertainment Networks somewhere out there. The antics of Fred, Ethel, Lucy, and Ricky may be how ETs think we interact with each other, and dropping atomic bombs on each other's cities may be how we settle our differences.

Common Internal Threats to Evolution and Survival of Intelligent Life on Planets

Following their initial formation, all planets are as hot or even hotter than the surfaces of their home stars. The universal physical laws of "thermodynamics" (i.e., conservation of energy) dictates that most of this heat must go somewhere else if life (at least any kind of life we know about) will have any chance of developing. In the early years of any planets existence, this heat must be controlled for biological life to have any chance of getting started and, in later years, it must be contained in order for life to survive.

Erupting volcanoes are more severe and frequent in early histories of planets In addition to threats from objects or forces that are located outside our world in outer space, such as collisions with asteroids, comets, neutron stars, black holes, rogue exoplanets, or even rare gamma ray bursts associated with exploding stars, other major extinction events could be triggered by internal factors located inside our own world from widespread volcanic outbreaks plus intense eruptions of isolated super volcanoes. All planets, including our own Earth, contain much more internal heat in the early years following their initial formation than they do in later years. Many exoplanets, and especially those located closer to their home stars in the so-called warmer habitable zones would, in the early years following their formation, be susceptible to "blowing their tops" so-to-speak via sudden surges in volcanic activity. Scientists believe that such outbreaks of severe volcanic activity were a major contributing factor in several of the mass extinction events that killed large numbers of our early ancestors (Rothery 2010). And, of course, many scientists believe that intense volcanism, while not being the major player in the death of the dinosaurs 65 million years ago was possibly a significant contributing factor. Therefore, it is likely that young worlds located elsewhere in the universe would also be vulnerable to these same kinds of internal heat related threats.

While volcanoes would be expected (like they were on our planet) to generally be far more frequent and more violent in the early years following their home planet's formation than in later years, they would most likely always exhibit a very wide size range. On our planet, although today's volcanoes, due partly to the reduced internal heating of the planet, occur more infrequently they can still be a major local threat to cities and human populations that are located nearby. One very rare but extremely violent form of volcano, which geologists refer to as a *super volcano* can, however, today still pose a significant global threat to life. Geologists who specialize in studying volcanoes (i.e., volcanologists) believe that at present possibly as many as seven such extreme forms of volcanoes exist worldwide. These super volcanoes are located in Yellowstone National Park, Wyoming, Long Valley, California, in the Jemez mountains of New Mexico, in Sumatra, Indonesia, in North Island, New Zealand, in Kagoshima, Japan, and only three miles below the ground in Naples, Italy. Super volcanoes, when they do erupt, produce explosions that are typically thousands of times more intense than those associated with normal volcanoes. Unlike normal volcanoes, super volcanoes occur when molten lava flows upwards from a hot spot at the mantle-core boundary for very long periods of time (thousands or millions of years) but encounters a ground surface that is unusually thick and hard which, unlike the softer ground that is located above normal volcanoes, prevents them from easily breaking through the ground to reach the surface of the Earth. A huge underground chamber filled with magma slowly builds over time and increases in pressure until eventually it explodes in a powerful eruption. The "good news" (if any can be associated with super volcanoes) is that, since these supermassive volcanoes are so slow in building themselves, they erupt very infrequently. The typical super volcano only erupts once every few thousand (or million) years. But when they do erupt, they typically produce a global level catastrophe. And, unfortunately, the rate of cooling of the core of our Earth is entirely too slow to allow super volcanoes to vanish as a major threat to life on our planet (or perhaps some other exoplanets) any time soon. ETs on some other worlds might also have to contend with similar ominous threats to their futures, unless they have the knowledge to implement some kind of technological intervention.

While no super volcanoes have erupted in over 70,000 years, the largest volcano eruption (Mount Tamboro) in Earth's recorded history, while considerably smaller than even the smallest of the six or seven earlier super volcanoes, occurred in April of 1815 in the Indonesian Ocean. This volcano erupted with a force estimated to have been somewhere between 100 and 150 times more powerful than the more recent Mt. St. Helen's eruption in Washington State in 1980. This volcanic eruption was so intense that it produced what historians call the *"year without a summer"*. The amount of dust, ashes, and other debris that was thrown into the upper atmosphere was sufficient to significantly block the sun's rays worldwide. This event caused a dramatic drop in average surface temperatures that was similar, but not nearly as intense, to what happened during the dinosaur extinction event. The sunlight blockage effect started in April 1815 and extended until the early part of 1817. For a large portion of the world the 1816 summer agricultural growing season did not happen. The summer of 1816 was wet and cold, and many of the summer crops totally failed.

On our own planet, another major super volcano is located beneath Yellowstone National Park in the United States. This volcano is believed to have erupted at least three times since Earth's formation. The three events were separated by an average period of 600,000 years. The last eruption occurred 640,000 years ago. Super volcanoes, unlike the more common varieties of smaller less severe volcanoes, are very difficult for geologists to identify. Unlike the traditional volcano which exhibits a tall mountain-like extrusion above ground with a hole at the top where hot magma and gases escape during eruptions, super volcanoes are not always visible above ground, i.e. do not exhibit tall volcanic type mountains. The land surrounding such volcanoes is typically flat or even slightly depressed except for the presence of smaller hot geysers (e.g., Yellowstone's Old Faithful geyser) and hot springs lakes. An extremely large magma chamber is, nevertheless, present a short distance below ground which provides the heat for the geysers and hot springs. When a super volcano erupts, the explosive ejection of the magma and gases from the underground magma chamber leaves a gigantic empty hole that is immediately filled by the collapse of the overlying land to form a huge and widespread depression. This depression is called a *caldera*. The caldera in the western United States that was created by the super volcano eruption that occurred 640,000 years ago is huge and measures approximately 32 miles by 45 miles. It is ironical that this huge caldera is remarkably flat (i.e., shows no significant signs of a depression) since at the time of the last eruption, a large mountain range was positioned above the magma chamber and when the land collapsed the mountains fell into and filled the depression. The fact that the last major eruption occurred 640,000 years ago, however, should not be interpreted to mean that we are now 40,000 years overdue for the next eruption and that everyone living west of the Mississippi river should immediately move to the east coast (or to Hawaii).While this volcano could erupt tomorrow, geologists tell us that the probability of a major eruption in the next 100 years is very small. Still, If this volcano were to explode at full force, it would devastate much of the continental United States and produce a worldwide catastrophe (freezing temperatures, destruction of vegetation-based food chains along with entire animal and plant ecosystems) that would last for many years. The destruction and human death toll in much of the continental United States would be severe, with almost total loss of life in a radius of several hundred miles from the center of the eruption. Although the eventual human death toll in most other parts of the world would be considerably less severe, mankind's technological civilization as we know it would be so severely damaged worldwide that it would require a very long time, perhaps many centuries to recover.

Finally, to emphasize the extreme threat to life posed by super volcanoes, I have attached three illustrations for the reader to see which demonstrate how much of a threat super volcanoes could be to the future of life on our planet as well as planets close in size to Earth elsewhere in our universe. Figure 6.13a shows an artist's view of the huge underground magma chamber that today lies directly below the site of the Yellowstone super volcano in the western United States, while Fig. 6.13b illustrates the absolutely incredible size of the ash field that was produced by the last two full eruptions of the Yellowstone super volcano circa 640,000 and 2 million years ago. Both Yellowstone eruptions were extremely severe and produced worldwide destruction.

However, since the last major eruption of a super volcano on our planet occurred approximately 70,000 years ago on a Lake named *Toba* in Indonesia, our scientists have no photographs of actual eruptions of super volcanoes. Figure 6.14a shows a photograph of the Pinatubo volcano eruption that occurred in the Philippines in 1991 which, while not a super volcano, provides the reader with a clear hint as to how large and destructive even the small non-super volcanoes can be. (If this volcano had occurred below Naples, Italy, the death toll would have been catastrophic for our Italian friends!) Finally, Fig. 6.14b shows a drawing by the USGS Comet program that depicts the relative sizes of six of the largest volcanoes (both normal and super volcano size) in Earth's recent history. The last Yellowstone super volcano eruption 640,000 years ago is estimated by geologists to have been 4,000 times more intense than that of the earlier "non-super" Mount St. Helen's eruption in 1980 (note size of the St. Helen's ash field in upper left corner of Fig. 6.13b). It is, therefore, quite possible that similar types of super volcanoes that are capable of global level destruction, could exist on other exoplanets in our universe. Our astronomers definitely see evidence of severe (past or present) volcanic activity on other planets (e.g., Mars, Venus) and even some of the moons (e.g., Jupiter's moon *Io*, Saturn's moons *Titan*, and *Enceladus*), and Neptune's moon *Triton* in our own solar system. Many scientists believe volcanic activity, even in the absence of tectonic plate activity, although destructive, could be important for the development of life elsewhere in our own solar system as well as other planetary systems since it provides a source of gaseous atmospheres as well as the extrusion of important life friendly chemicals from deep below the moon's or planet's surface.

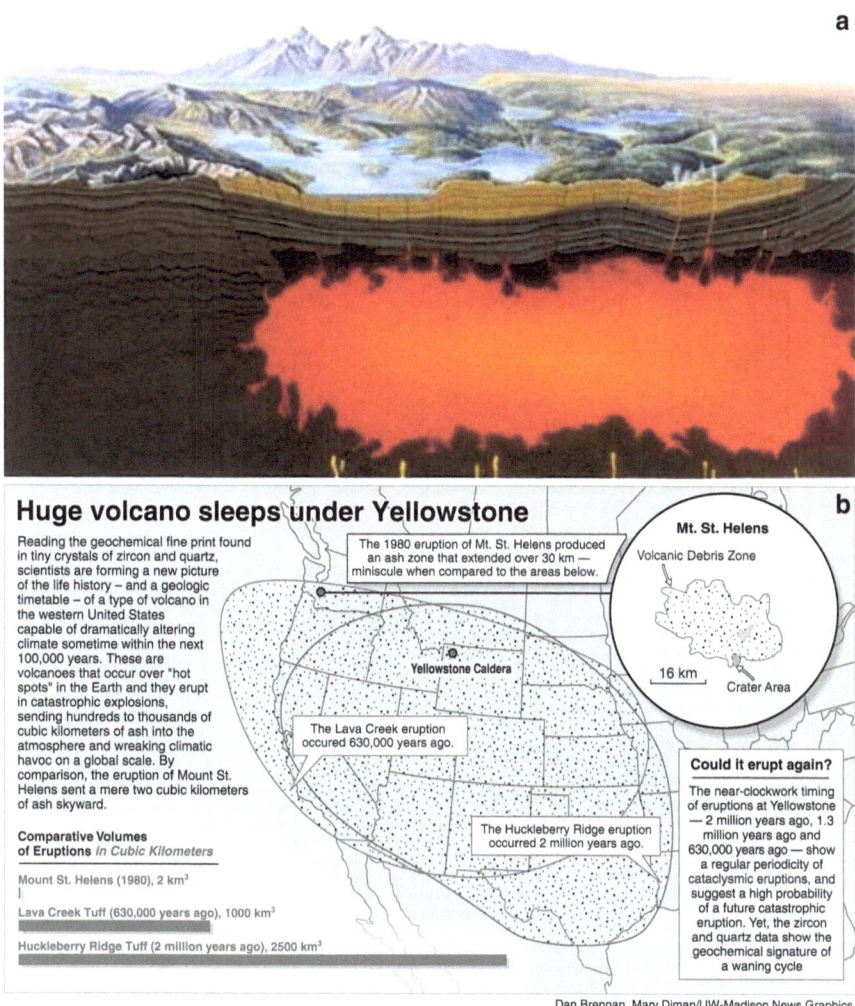

Fig. 6.13 (**a**) Shows an artist's drawing of the extremely large magma chamber that is located beneath the super volcano in Yellowstone National Park (image credit: United States National Parks Service), while (**b**) depicts the incredible size of the ash fields that resulted from the last two major eruptions of this super volcano approximately 640,000 and 2,000,000 years ago (image credit: USGS). It is worth noting that the caldera (i.e., slight ground depression) left over from the last super volcanic eruption is huge and, although oval shaped, measures roughly 32 by 45 miles. In drawing "**a**" this caldera is shown as a *lighter brown colored area* below the ground surface that is remarkably similar in size (length/width), but not depth, to the magma chamber that lies below it

Severe earthquakes may also threaten inhabitants of other planets As we will see later in this paragraph, volcanoes and earthquakes are not independent threats to life on planets. Any planet that hosts a surface in which the outer crust is not one continuous entity but is made up of many separate "**tectonic plates**" that lie

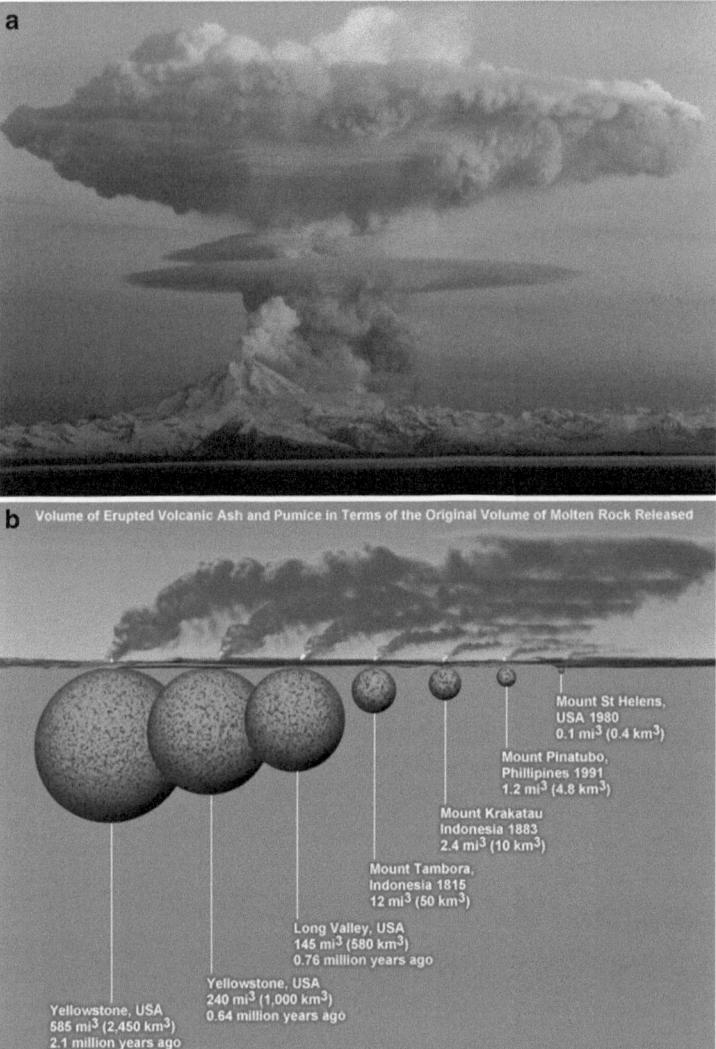

Fig. 6.14 While we have no photographs of any super volcano eruptions to show the reader how powerful this form of volcano can be, we can show (**a**) a photograph of the 1991 eruption of the Mount Pinatubo volcano in the Philippines which, while not a super volcano, is the largest volcano eruption to have ever been photographed which was even larger than the earlier 1980 Mount St. Helens eruption that occurred in Washington state in 1980. To depict how large and powerful volcanic eruptions can be, the relative amounts of volcanic ash that were expelled by the last seven major volcano eruptions in Earth's history are shown in (**b**). The two images on the *left side* of this drawing are those of the most recent eruptions of the Yellowstone super volcano. A *growing number of scientists (geologists who specialize in the field of "volcanology") have, in the last few years, started to believe that the Yellowstone super volcano is continuing to grow in size and may produce another massive eruption in the not too distant future.* A supermassive explosion of this volcano would produce a worldwide disaster that could definitely be bad for life plus catastrophic for our human civilization. If the Mount Tambora eruption of 1815 in Indonesia pumped enough ash into the atmosphere to produce a worldwide agricultural crisis or "*year without a summer*", what would the next Yellowstone eruption, which might be as much as 60 times larger, do to our world? © The COMET Program/USGS (image credits: USGS)

adjacent to each other and which move independently of each other, are susceptible to catastrophes associated with both volcanoes and earthquakes (Rothery 2010). Volcanoes and earthquakes are the direct result of tectonic plates being shoved around (and colliding with or sliding underneath each other) in response to heat rising from the planet's hot interior (core). For thousands of years, mankind's only plausible explanation for why we occasionally experience sudden unexpected and frequently deadly Earth tremors or "ground shakings" was that some higher deity (or deities) was angry with us. Ironically, it took a world war for man to discover the real answer to this puzzle. Following world war II, the military element of our society (in all its wisdom) decided that in order to make the future of submarine warfare more effective, we needed to totally map all of the underwater surfaces that lay below our seas and oceans. Prior to this time, our geologists believed that the outer surface or crust of the entire Earth was one continuous structure. When they started mapping the ocean bottoms, the geologists now suddenly discovered that *the crust of the Earth is not a single continuous structure that extends around the world but is separated into at least 14 distinct plates of different sizes that lie adjacent to each other*. These different crustal structures, which geologists now call *tectonic plates* (Fig. 6.15a, b) are separated from each other by cracks or discontinuities that allow them to move independently of each other.[9]

One of the strangest and most pervasive facts of life on our planet, and perhaps also on many other planets in the universe, is that sometimes *the "bad" comes with the "good"!* For many years, our geologists have been aware of the fact that, in addition to being located in our sun's warm habitable zone, the major reason that our planet has enjoyed a temperate and life friendly environment for so long is that the co-evolution of life and geology on our planet has created some unique geological phenomena or processes that are critical to allow our carbon-based life forms to keep on reproducing. However, in doing so, these same geological processes can and frequently do kill many of us. In a rather weird but quite accurate sense, all life on our planet is totally dependent on those things that volcanoes and earthquakes do other than create mayhem in our lives. If volcanoes did not erupt and spew fire and lava, or earthquakes did not destroy lives and property, our planet's thermostat system for maintaining a relatively pleasant climate (i.e., swinging back from "not too hot" in the summers to "not too cold" in the winters) would, along with life, go away. And, if tectonic plates did not exist (or did not move) carbon and other life sustaining elements or nutrients would not be able to be recycled back and forth from inside the Earth to maintain a constant supply of fresh new construction materials for creating new life.

Some scientists believe that the Earth may presently be the only planet in our solar system that hosts tectonic plates. Since the specific details of how tectonic

---

[9] The reader may want to look at the author's earlier book (Cranford 2011), where I present a detailed discussion of the importance of tectonic plates on our planet for allowing the transfer of heat from the core of the Earth to outer space as well as the recycling of life critical elements (e.g., carbon) for purposes of supporting a long term life friendly environment.

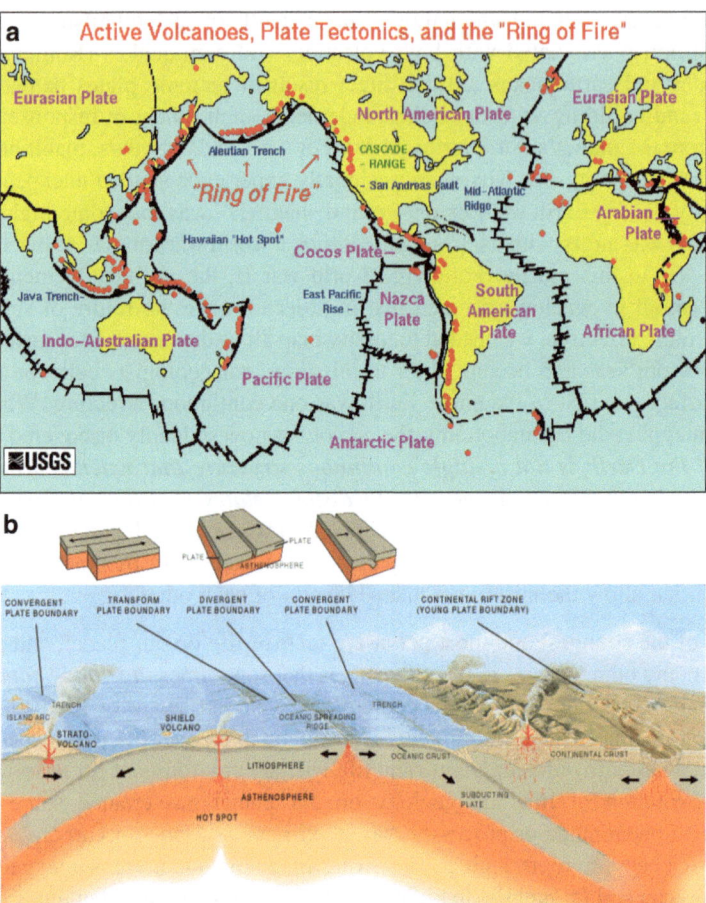

Fig. 6.15 (**a**) The Earth's outer crust is not one continuous structure that covers the entire planet. Shortly after WWII geologists discovered it consists of at least 14 separate plates that lie adjacent to each other which move independently of each other (image credit: United States Geological Survey). Topinka USGSICVO, 1997, modified from: Tilling et al. (1987) and Hamilton (1976). (**b**) Shows that, as heat rises from the very hot and semi-fluid core and mantle of the Earth to the rigid outer surface or *oceanic crust* of the planet it causes the tectonic plates to move relative to each other. When the rising heat makes the tectonic plates move, they can move relative to each other in three different ways. The *top left drawing* shows two plates suddenly sliding parallel to each other (i.e., they "side swipe" each other). When this happens, they frequently produce severe earthquakes such as those associated with the 1906 San Francisco earthquake. When two plates (*upper right drawing*) move toward each other, and one plate is heavier than the other, the heavier plate will move (i.e., subduct) below the lighter plate and create frictional heat that can rise to the surface of the Earth to produce volcanoes. Thousands of these volcanoes are found all over the world. The most severe currently active of these volcanoes are shown (in *red*) and are collectively referred to as the infamous "**Ring of Fire**". And, finally, if the two plates (*middle drawing*) that are moving toward each other are equally heavy they will collide and, and having nowhere to go, will push the ground upward to produce mountain chains. And, if these two equally heavy plates are moving away from each other they will produce what is known as a "continental rift zone" which resembles a long deep valley or depression in the ground (image credits: Wikipedia Commons)

plates came about on our planet is presently not known, there is the possibility that two other members of our solar system's inner rocky planet group, e.g., Venus and Mars, may have had tectonic plates in their earlier histories which, as they did for us, may have assisted in jumpstarting life. Growing numbers of scientists are now strongly suggesting that it is the unique combination of tectonic plates, volcanoes, location (habitable zone), plus our planet's size and ability to host surface water resources that explains why our type of life is here today. Many scientists also believe that the presence of a relatively large moon may have acted to stabilize the daily spin (day/night) cycle as well as the rotational axis of our planet as it orbited the sun. This orbital stabilization factor, in addition to allowing predictable and stable periods of warmer and cooler climates that are critical to supporting our types of seasonal plant life, would have also kept our oceans constantly "churning" (producing tides) which is absolutely critical to moving life critical nutrients and other materials from place to place plus also allowing land and sea species easy access to each other to allow further diversification of life on our world.

Many scientists also believe that, while primitive microbial forms of life may be common in the universe, it is possible that twins of Earth that can host advanced multicellular life forms like man may be extremely rare. Although the planet hunters are indeed beginning to find evidence that twin Earths may be rare in the universe, what is currently totally unknown is whether the chemistry and mechanics of the evolution of life (or whatever life is) may be limited or restricted to a limited set of conditions, or whether the evolution of life has few boundaries and can easily adjust to virtually anything nature throws at it. The recent discovery by radio astronomers that the chemical stuff or precursors for our kind of Earth-bound carbon-based life may be relatively easy to manufacture in remote regions of space and deliver to appropriate planets or moons inside meteors or comets to jumpstart life now also suggests the possibility that our form of life may not be all that rare out there but may actually be more the norm than the exception in the universe. I personally am very frustrated that I have no way of stopping or slowing my biological aging clock so that I can be here to witness the outcome of this, the greatest of all of life's many mysteries. Just how diverse is this thing we call life in the universe, and just how rare or "exotic" is our type of life out there?

Dramatic shifts between extreme hot and cold climates may be common in early histories of some planets As mentioned above, many of the common internal threats to life's survival (e.g., intense volcanic outbursts, or freezing temperatures caused by blockage of sunlight by debris clouds thrown into the atmosphere by volcano eruptions or asteroid impact events) are themselves secondary to preceding external events. Still, early in their post-accretion histories, many exoplanets are likely to be internally unstable due to excessive and relatively poorly controlled geophysical and climatic-related energy sources (Courtillot and McClinton 1999). Our own Earth was, during the first few billion years of its existence, a seething hotbed of violent volcanic activity and outgassing of various kinds of gases from its interior that had been planted there by the intense celestial pounding that the Earth experienced during its formation and subsequent heavy bombardment periods. Our own solar system also had to deal with an earlier cooler sun that made it difficult to establish a long term stable habitability zone. The sun, even today, is continuing to get hotter, which means that this instability will continue on into mankind's future (or whatever life forms may replace us). Most stars in the universe share this same tendency to "heat up" as they age. Therefore, episodes of having to tolerate excess climatic heat alternating with ice-age cold spells were probably a virtual way of life for our earliest primitive ancestors, and could be for extraterrestrials elsewhere in the universe. Our scientists tell us that, sometime around 650 million years ago, huge numbers of our early ancestors froze to death when the Earth's CO_2 thermostat control system allowed the Earth's temperature to plunge to the point that glaciers from both polar regions may have reached almost as far as the equator (Fig. 6.16a, b). This catastrophe created all kinds of empty environmental niches in the world's oceans that may have left the door wide open for the evolution of more complex multi-cellular creatures later in life's evolutionary saga (Walker 2003). Similar *snowball Earth* phenomena might occur on other exoplanets in the early years following their formation. While the severe snowball Earth event 650 million years ago may ironically have acted to facilitate and speed up the development of later species of more complex multicellular life forms, the biggest threat to the development and survival of life on our planet was and still is, even today, the eruption of supermassive volcanoes (and, at least on rocky worlds like ours, the volcanoes' "partners in crime" known as severe earthquakes and tsunamis). Historically, in the last 300 million years, the Earth has experienced at least three major extinction events that were due mostly or entirely to outbreaks of severe volcanic activity that inflicted heavy damage to life (see Fig. 6.17 plus also Figs. 7.1 and 7.2 in Chap. 7 for additional information on the geological and life timelines of our planet). The most recent extinction event occurred approximately 65 million years ago (mya) at the end of the Cretaceous Period. This event occurred in west-central India in a place referred to as the "Deccan Traps" in the Indian Deccan province. This outbreak was caused by severe volcanic eruptions that may have lasted for as long as 30,000 years. This volcanic outbreak killed large numbers of dinosaurs (Fig. 6.18), (*which definitely added to the dinosaur death toll inflicted by the infamous End-Cretaceous or K-T asteroid catastrophe that hit almost at the same time on the other side of the world on the Yucatan peninsula in the gulf of Mexico; see Figs. 6.2 and 6.3*). The other two severe outbreaks of volcanic eruptions occurred about 200 billion years ago (End-Triassic or "End-Tr" event) at the end of the Triassic Period, and 250 million years ago at the end of the Permian Period. This last series of eruptions (End-Permian or "End-P" event as geologists label it), was by far the most severe mass extinction event in history. It occurred almost continually over an incredible time period between about 300 and 250 million years ago. Geologists have appropriately labeled this extreme event as the **Great Dying event**, since many Earth and life scientists believe this catastrophe came very close to actually destroying all life on our planet (Erwin 2006). It was virtually a worldwide event affecting both oceans and land areas over huge regions of the globe, in which huge numbers of our ancestors were boiled or burned to death. Over 70 % of all land life (Fig. 6.19a)

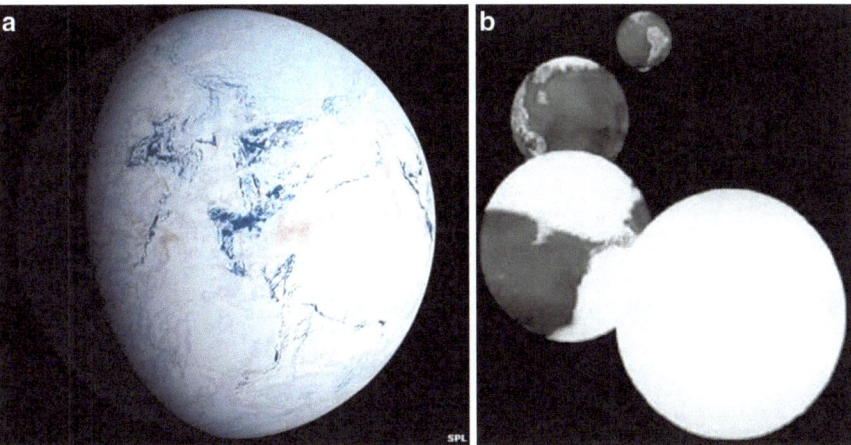

Fig. 6.16 Although our planet usually enjoys a life friendly temperate climate, historically there have been periods of extreme cold (circa 650 million years ago). (**a**) Shows a series of images that depicts how the ice and snow coverage started at the north and south poles and then gradually extended toward the equator to engulf the planet, while (**b**) depicts the maximum ice and snow coverage that many scientists believed occurred which may have included the entire Earth (image credit: NASA/SPL)

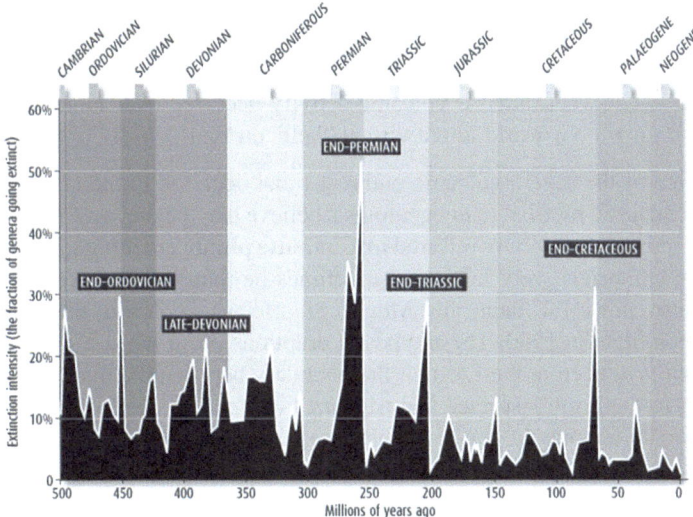

Fig. 6.17 Shows a diagram depicting the severity of the five major mass extinction events that occurred on our planet (image credit: Wikipedia Commons; see also, *Figs. 7.1 and 7.2 in Chap. 7* for additional important information on the evolution of geology and life on our planet)

Fig. 6.18 Shows an artist's drawing of the incredible devastation, including the killing of large numbers of dinosaurs, that resulted from severe volcanic eruptions (End-K event, or end "Cretaceous" event) in the Deccan Traps region of what is now south-central India. This disaster began about the same time (65 million years ago) as the Mt. Everest sized asteroid was impacting what is now northern Mexico. This Deccan Traps disaster involved a series of intense mantle plume volcanic eruptions that covered a 193,000 square mile area and may have possibly lasted 30,000 years (image credit: Artist illustration by NSF based on research by Gerta Keller and collaborators in India)

and an astounding 97 % of ocean life (Fig. 6.19b) on our planet perished in this extinction event when intense volcanic activity occurred over huge areas of our planet that turned our world into a virtual "hell" on Earth.

All three of the severe volcanic outbreaks that occurred in the last 300 million years were caused by what many geologists believe may be the most severe form of volcanic activity, commonly referred to as **mantle plume eruptions**. Mantle plume eruptions, while relatively rare, can sometimes be thousands or even millions of times more powerful than the Mount St. Helen's volcano that erupted in Washington State in 1980. These types of eruptions occur when a gigantic ball of hot molten lava deep in the Earth at the boundary between the mantle and molten core rises to the Earth's surface and triggers a continuing series of eruptions from hundreds or thousands of separate volcanoes that are scattered over a large geographical region.[10] This particular volcanic event is believed to have resulted in an

---

[10] It is the constant movement of tectonic plates in the Earth's crust that makes mantle plume eruptions so dangerous. While rising lava plumes tend to stay in one constant position over incredible lengths of time (thousands or more years), the ground (i.e. tectonic plates) located above the rising lava plumes continue to slowly move (very slowly, perhaps only a few inches per

Fig. 6.19 Shows artists drawings of the "*mother of all*" of our planet's known mass extinctions that many scientists believe came close to actually ending all life on our planet. This extinction event, which is known as "*The Great Dying Event*" because of both its length and severity is believed to have occurred almost continually from about 250 to a little over 300 million years ago, and was incredibly extensive, involving both land and water areas virtually worldwide. (**a**) Shows an artist's drawing of the destruction that occurred to much of the world's land areas (image credit: NASA Lunar and Planetary Institute), while (**b**) shows that virtually all of the world's sea life was also destroyed during this horrific geological catastrophe at the end of the Permian period (End-P) (image credit: Karen Carr of Karen Carr Art Studios)

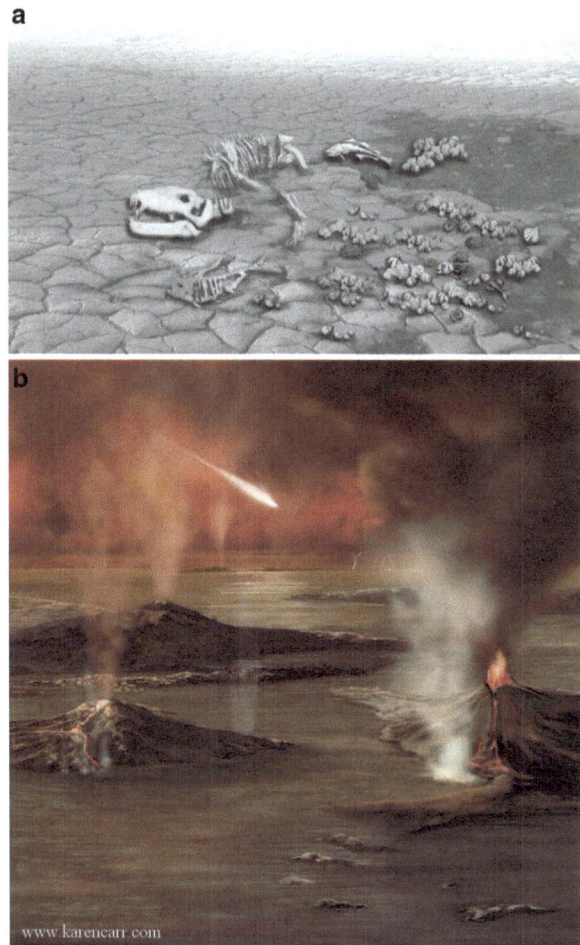

area about the size of the modern continental United States being covered in many locations by the equivalent of 1,000 ft. deep (or greater) lava flows. To make things even worse, some scientists now estimate that these mantle plume volcanic eruptions may have occurred repeatedly over a period as long as 10 million years. So, on our own planet, life of any kind, and especially intelligent life, is extremely fortunate to still be here today. How lucky might our extraterrestrial neighbors be?

year). Over a long period involving many years a single lava plume will continue rising and repeatedly "punching" through the ground to produce new volcanoes and new land. It was the occurrence of large continual rising plumes of lava from deep in the Earth's core that actually built the long Hawaiian Island chain as well as other island chains elsewhere in the world. Although tectonic plates did not move any faster during the Permian Period, there were more mantle plumes present than there are today. This is why virtually the entire planet was plagued by many more deadly volcanoes during this early period in Earth's history.

So far in the present chapter, I have focused on the major types of natural catastrophes that originate from sources located external or internal to planets that could threaten the evolution of life on planets like our Earth. However, as I emphasized earlier in this chapter, our planet hunting astrobiologists have started finding that planetary systems that are close to being twins of our home planet or our own solar system appear to be relatively uncommon. Our best ground-based telescope (interferometer) systems plus NASA's powerful new Kepler space telescope now indicate that a large proportion of the stellar systems that we have been able to identify in other parts of our galaxy in the last few years appear to be non-Earthlike and, therefore, probably unfriendly to anything resembling our own planet's familiar carbon and water based forms of life. Some of the exoplanets discovered so far are huge (even larger than Jupiter and Saturn) and extremely hot due to their orbiting very close to their home stars, or frozen balls of gas orbiting well beyond the habitable or goldilocks zones of their own planetary systems.

On February 1, 2011, the Kepler telescope team released a *graphic summary of the estimated size, orbital periods, and surface temperatures of the first 1,202 exoplanets that were discovered in the first year of active exoplanet searching* (Fig. 6.20). These two graphs strongly make the case that most of the current cadre of potential exoplanets discovered with the Kepler space telescope are definitely not twin Earths. From 1995 (when Swiss astronomers discovered the first exoplanet) until close to 2010, it seemed that the only exoplanets our astronomers could easily detect were "hot" super-Jupiter gas giant planets that were making their home stars produce "big" and "fast" wobbles as they orbited them. With the launching of the Kepler space telescope in 2009, the planet hunters could now use more sensitive techniques involving the detection of even smaller brightness changes produced by transiting exoplanets plus eliminate fuzzy images caused by our dirty and turbulent atmosphere. While the Kepler scientists are still finding giant planets, some of which still orbit close to their home stars, they are now also being able to observe smaller and smaller exoplanets, and are now beginning to find exoplanets as small as Mars and Mercury.

Some of the planets being discovered by Kepler can only be described as being "monsters from hell" with rocky surfaces plus extremely hot (several thousand degrees centigrade) atmospheres that host storm-like conditions that would make the Earth's strongest hurricanes look like a gentle summer breeze in contrast. Some systems have even been found in which the orbiting exoplanets, instead of being bathed by a relatively friendly and warm sun like we enjoy here on Earth, are circling huge giant stars or the leftover remains of supernova events that are constantly bombarding them with horrendously deadly radiation that would instantly kill any kind of life that we could imagine. Yet, some of our diehard astrobiologists now believe that, based on what some of our own extremophile relatives on Earth are teaching us, life of some kind might still be possible, even on some of the more hostile of these so-called other worlds.

Thus, solar systems like ours definitely now appear to be more the exception than the norm in the universe. If, as many scientists now believe is entirely possible, life is very flexible and resilient and is capable of popping up in the most extreme

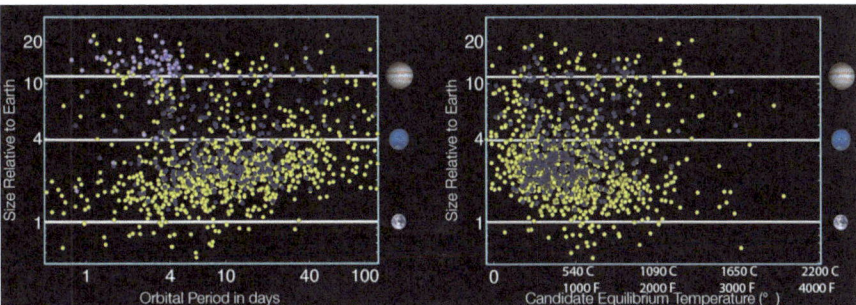

Fig. 6.20 On February 1, 2011, the Kepler telescope team released a *graphic summary of the estimated size, orbital periods, and surface temperatures of the first 1,202 exoplanets that were discovered in the first 11 months of formal exoplanet searching.* In both graphs, the *vertical axis* shows the size of the different exoplanets relative to the sizes of Earth (*bottom*), Neptune (*middle*), and Jupiter (*top*). While the *horizontal axis* of the *left graph* displays the orbital periods (length of 1 year) of the newly discovered exoplanets, the *horizontal axis* of the *right graph* depicts the estimated average surface temperature of the different planets. While some of the planets are circling sun-like or larger stars, others are circling smaller cooler stars known as red dwarfs. Some exoplanets in these red dwarf systems have much shorter orbital periods than the Earth but since their home star is significantly cooler than our sun, they may still be orbiting in cooler habitable zones (goldilocks zones) very similar to what we enjoy on Earth. However, as can be seen in the *right graph*, some of the smaller exoplanets are still orbiting in very hot zones, as are many of the larger planets. In both the *left* and *right graphs, the* purple colored *entries represent exoplanets discovered before Kepler, i.e., prior to 2009. The* blue entries *represent exoplanets discovered by Kepler as of 2010, while the* yellow entries *depict exoplanets discovered as of 2011 by Kepler.* It does seem, however, that switching to more sensitive search techniques (e.g., the transit method) for detecting smaller exoplanets has not (yet) eliminated the incredibly large range of differences that we are seeing in the size, temperatures, and whereabouts (orbits) of other exoplanets in the universe. As of the writing of this book (August, 2013), stellar systems harboring "twin Earth" type exoplanets continue to be nature's oddball (image credits: Kepler/Nasa.gov)

environments, the variety of threats I listed earlier in this chapter may be only the beginning of a potentially much longer list of potential threats that ETs on other worlds might have to contend with in order to survive. Life forms living on many other worlds may be faced with entirely different kinds of threats that our scientists presently know nothing about. Threats from close by exploding stars, or intense earthquakes and volcanoes would probably also occur on many other planets, especially if they are small and wet enough to support plate tectonics which appears to be critical (at least as it was on Earth) to allowing temperate climates and the constant recycling of life sustaining chemicals such as carbon and nitrogen. The identification of other even stranger internal or external forms of threats to the development of life will have to await future investigations by our astrobiologists as they, or their remote sensing technologies, begin to explore far distant regions of our universe. One thing is sure—our future astrobiologists and space explorers may encounter many surprises in their daily work. Life out there may be so different from life as we know it on Earth that the whole concept of what is and is not alive will have to be totally redefined over and over again by future generations of scientists. However, whether or not mankind will be able to survive long enough to allow future generations of space explorers to make these profound discoveries is, unfortunately, not yet settled. It is now time for the author to turn to a discussion of what he unfortunately believes may be mankind's most serious threat to life on our planet, and that is not geology, but man **HIMSELF!!**

Correlations Between Evolution of Intelligence and Rise of Technologies That Can Destroy Life

It unfortunately does seem that the smarter man has gotten over the years, the more of a threat he has become to himself as well as to the environment in which he lives. Primitive man first invented clubs, then spears and bows and arrows to kill both his food as well as his human adversaries. Then came guns and cannons, and now we have weapons armed and deployed all over our planet that are powerful enough that, if launched into action, could destroy all human life on our planet in just a few days (as the world came dangerously close to experiencing in October, 1962, during the infamous "Cuban missile crisis"). In a very real sense, man's intelligence has become both his greatest potential blessing as well as the greatest threat to his future on our small fragile planet. The rise of intelligence in our species has brought with it the "good", the "bad" and the "ugly" with respect to our future and, in man's case, the word "ugly" appears at times to be unfortunately synonymous with the word "stupidity"!

"Kill or be killed" may be a universal feature of life's earliest beginnings In earlier chapters, the author described how, at least on our planet, the biological evolution of intelligent life was not only incredibly slow but has, so far, failed to produce any nervous systems that are totally precise and accurate in terms of their functions. The brains of even the so-called most intelligent creatures on Earth, i.e., humans, are prone to making mistakes or errors. In addition, humans are the direct descendents of earlier predatory animal species that were forced to hunt, kill, and eat each other in order to survive.[11] In the jungle, predators spend much of their

---

[11] All biological systems need a readily available source of energy in order to survive and reproduce. Biological systems that perform more complex functions require more energy. On our planet, the reason that predation was needed to foster at least the early stages of the evolution of intelligent animal species is that the function we call intelligence demands more energy to work. Our nervous systems need to have a source of energy that is both quick and easy to acquire as well as being highly efficient (i.e., "more bang for the buck"). Herbivores (cows and horses) spend much of their time eating grass or other plants that have low energy content. Since these creatures do not need to be real smart to survive, they function quite well with low energy foods. Humans need to consume foods that are highly packed with energy such as meat and protein to keep their much more active brains going at full speed. This is why man's brain, which constitutes only 2 % of the total weight of our bodies, requires 15 % of our blood supply, 20 % of our oxygen intake, and 25 % of all the energy (glucose) we take in from feeding. Intelligent ETs on other worlds would

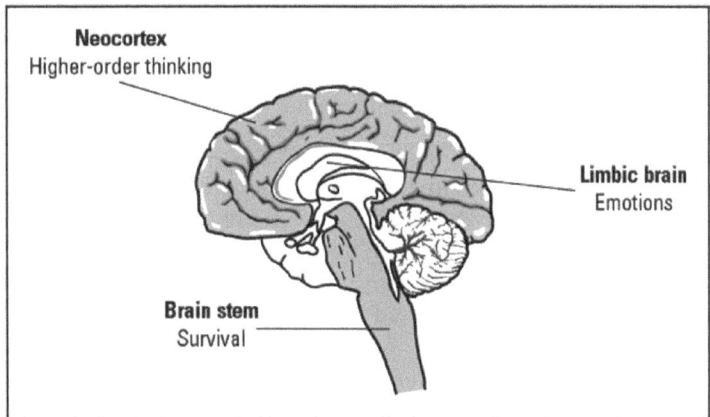

Fig. 6.21 Shows the size and location of the neocortex and limbic system components of man's brain which controls our intellectual and emotional natures, respectively (image credit: Laura Erlauer, Brain-Compatible Classroom, Association for Supervision and Classroom Development, 2003)

time looking for other animals to kill and eat or in taking evasive actions to avoid being eaten by other predators. The law of the jungle that only the fittest survive forced nature to implement evolutionary strategies that favored those creatures that had the stronger aggressive tendencies and the more effective cunning behaviors when it came to effectively and efficiently finding and killing their victims, plus successfully avoiding being eaten. This predatory lifestyle was undoubtedly the single greatest factor that triggered the rise of the behavioral attribute called intelligence in the different animal species. In a very real sense, intelligence and aggressiveness are strong inter-dependent consequences of early man's (and other animals) being forced to pursue a predatory lifestyle in order to survive.

Scientists (e.g., Sagan 1986; MacLean 1990) tell us that the brains of modern man (Fig. 6.21) and the brains of all modern day predatory species (e.g., snakes, alligators, sharks, lions, bears, etc.) share similar complex neuronal circuits (i.e., the so-called sub-cortical *limbic* cortex regions of the brain) that control their owners emotional states (anger, fear, love, hate, competitiveness, aggressiveness, etc.). While the limbic area of the brain is a very primitive structure that has been present in all animals from the earliest amphibians and reptiles up to man, there is another region of the brain that is unique to man's biological line of evolution. This brain region, which is known as the *neocortex*, is located on top of the sub-cortical limbic area of the brain. In mammals, this cortical area has steadily grown larger and larger (see Fig. 3.2b in Chap. 3) in relation to other areas of the brain (including the limbic area). In the primate line of evolution, and especially in apes and humans, this

---

also likely have "energy greedy" nervous systems that would require they engage in some form of predatory lifestyle or its technologically developed equivalent.

cortical area has grown quite large. Scientists believe the neocortical part of the brain houses the complex neural circuitry that controls higher level intelligence. Thus, while man is supposedly the most intelligent life form on planet Earth, he still possesses a very active sub-cortical brain region, the limbic area, that controls his emotions, and especially his competitive and aggressive nature.

From predation to intelligence to self-destruction?   Therefore, man as well as all other mammalian species, possess two distinct areas of the brain, one of which, the neocortex, controls intelligence and the other, the sub-cortical limbic areas, that controls our emotions. However, there is now considerable evidence that this evolutionary history may have created some kind of conflict or paradox in our species. Being intelligent and being aggressive are both good things if they are strictly used to survive in the wild, as was the case for our early ancestors. Nevertheless, when mankind transitioned from a predatory jungle lifestyle to a cooperative civilization form of lifestyle, man's intelligent and aggressive natures may have started to conflict rather than reinforce each other. In a cooperative civilized world where it was now possible to share the workload with other tribal members and not have to continuously engage in hunting, killing, running, and hiding to survive from day-to-day, some individuals could leave the basic routine chores (gathering food, tending gardens, etc.) to others and engage themselves in inventing new and better tools to make life more fun and comfortable for everybody as well as to protect the home team (tribe) from the aggressive assaults of compet-ing tribes. Man started to build weapons, first clubs, then bows and arrows, then guns, and eventually nuclear weapons. In the meantime, man's brain still housed the same old neuronal structures that had facilitated his survival for millions of years. Now, when his sub-cortical limbic system threw him into an aggressive emotional state, rather than killing a rabbit and eating it, he might push a button to unleash rockets carrying nuclear warheads to kill his tribe's enemies on the other side of the world. In addition to the possibility of terminating all life on our planet via the unleashing of nuclear weapons, man's compulsive intelligent nature has, since the end of World War II, also provided him the means by which he could inadvertently or deliberately end all life on our planet via biological weapons, or the accidental production and unleashing of chemicals into the environment that could destroy life on our planet. These nuclear and biological threats may also exist on many other exoplanets in the universe where life has managed to survive long enough and get smart enough to allow them to occur.

Development of technologies that can kill intelligent life as well as destroy environments   In addition to allowing us to build bombs and chemical agents that could destroy both us and our enemies, mankind's unique intelligent nature may also be leading us to a completely different kind of crisis involving our abuse of the environment in which we live (Ward and Brownlee 2003; Ward 2009).[12] In just the

---

[12] Another profound truism related to the rise of life (whether intelligent or not) on any planet anywhere in the universe may be that life, by its inherent nature, not only competes with other life

last few years, increasing numbers of our scientists have begun to believe that we are currently facing a possible *global warming* **crisis** that has the potential to possibly trigger a runaway greenhouse effect that could destroy all life on our planet, or at least make our environment totally hostile (Ward 2007). Ever since primitive man learned to burn wood using fire, he has been contributing to the creation of carbon dioxide (CO_2), which is the predominant form of greenhouse gas in Earth's atmosphere. For many years, our scientists have been well aware of the fact that the presence of atmospheric greenhouse gases is the major reason life was able to develop on our planet. Without CO_2, our planet would be far too cold to allow the existence of liquid water on its surface which is critical to us carbon-based life forms. For thousands of years following his discovery of fire, the total amount of CO_2 that man created was virtually nil in comparison to other natural sources of CO_2 (volcanoes, forest fires, the respiratory waste-product of man and other animals, rotting of dead plant and animal life, etc.). However, beginning with the so-called Industrial Revolution in the mid-1700s, this situation changed drastically. Man began burning not only wood, but other forms of fossil fuels (coal, oil, natural gas), but this time to *build, build*, and not just cook and keep warm. Man started building large homes and even larger factories, all of which demanded huge amounts of energy. He quickly discovered that wood and fossil fuels were the easiest and quickest source of readily available energy. Once the Industrial Revolution started it quickly grew and within a short time many major cities began hosting smog-shrouded atmospheres that now contained levels of CO_2 gases that were suddenly no longer life friendly but were now quickly becoming a threat to life.

The atmospheric greenhouse effect itself was not discovered until 1824 by Joseph Fourier, and it was not until 1896 that another scientist named Seante Arrhenius discovered that CO_2 was the major gas that was responsible for creating this effect. By that time, the Industrial Revolution was in full swing in every major industrialized country in the world. By the late 1800s man was able to begin accurately measuring and keeping records of daily surface temperatures all over the world, and also precisely measure the relative amounts of CO_2 in the atmosphere as well as the amounts of CO_2 being pumped into the skies from his smoke stacks (plus in a few more years by his combustion engines). However, it was not until well into the twentieth century that man began to wake up to the possibility that the combination of industrialization, CO_2, and the greenhouse effect might not make for such a good mix with respect to man's future.

Starting in the second half of the twentieth century and continuing up to the present day (2014), some scientists and concerned citizens began to notice a number of changes in the world's climate and environment that gave them cause

---

forms to determine which particular species (whether animal or plant) gets to dominate the local environment, but also tends to take more resources from its environment than its environment can replace, plus also adds chemical pollutants or waste products to the environment that may eventually destroy it (see Ward 2009).

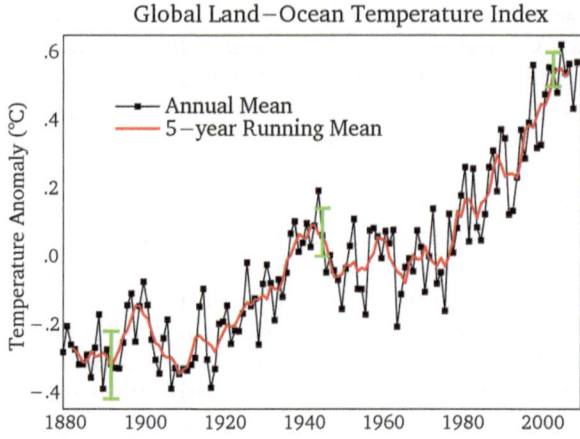

Fig. 6.22 Illustrates the dramatic rise in average atmospheric temperatures that has occurred worldwide since about 1900 (image credit: NASA)

for concern. The first major anomaly was the observation that starting around 1900, the global mean surface temperature started rising at a surprisingly fast rate. In the 100 year period between 1905 and 2005, the average world surface temperature increased by an average of 0.9 °C or 1.6 °F (Fig. 6.22). While an average temperature increase of just over 1.6 °F in one century does not sound like much, climatologists tell us that, in terms of the relative magnitudes of worldwide geophysical events, it is *huge*. During this same period, the amount of CO_2 in the atmosphere rose by a total of 35 %. The CO_2 level in the atmosphere measured in 2005 is the highest it has been in the past 650,000 years (Fig. 6.23). Still, what is most alarming is the fact that, at the present rate of increase, the world's average surface temperature could rise by as much as another 7.2 °F (4 °C) by the end of the twenty-first century. What would be the major consequences of such a worldwide temperature increase? Climatologists tell us that, while the effects would be widespread and generally severe, it would be very difficult to draw any kind of accurate predictions of how specific effects would vary in different locations around the Earth. If severe global warming becomes a reality in the next century, we will likely not be able to easily draw definite cause-and-effect relationships between specific weather events and the global warming phenomenon. Most regions of the Earth would suffer while others might, at least initially, actually benefit. Regional changes in the amount and pattern of precipitation may completely redo the map of the world in terms of flooding and drought zones. Both Africa and Western Europe would probably be thrown into severe drought conditions while Canada, due to a climatic warming trend caused by global warming might actually incur an agricultural "boom" due to warmer and wetter conditions and become the world's new "bread basket", at least for awhile. The snow covered winter zones of the Earth would drastically shrink, glaciers everywhere would melt at a rapid rate and the North Pole's ice cover would disappear and Antarctica would eventually become a dry wasteland (or an inland ocean due to rising sea levels). The loss of snow and ice worldwide would further decrease the amount of sunlight that could be reflected back into space. Of course the fact that

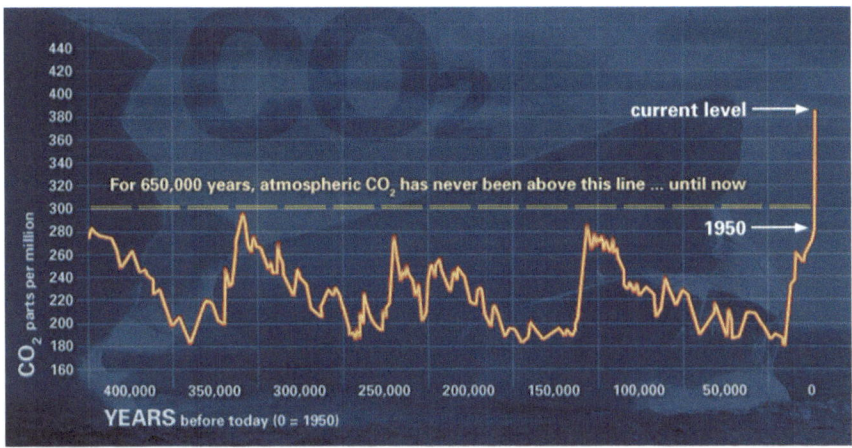

Fig. 6.23 Since 1950, the average concentration of carbon dioxide (CO_2) gases in Earth's atmosphere has suddenly risen to its highest level in over 650,000 years, suggesting we may now be experiencing global warming (image credit: NASA)

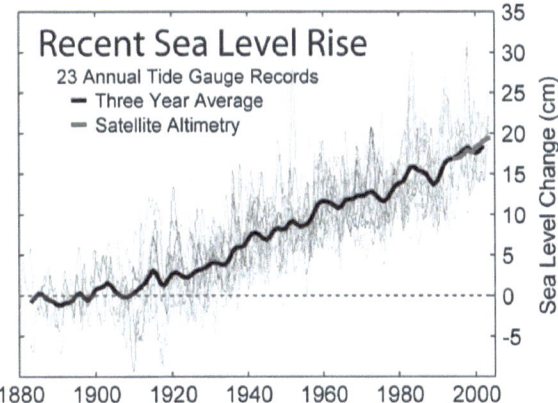

Fig. 6.24 The recent global warming trend has started causing snow and ice covered regions of the world to begin melting which is rapidly producing a rise in average sea levels which will begin to threaten seashore cities and populations (e.g. Holland, Miami, and the author's own home town in New Orleans) by the beginning of the next century (image credit: Wikipedia Commons)

ice and snow areas worldwide would start melting and shrinking would produce rises in sea levels everywhere. In fact, such dramatic rises in sea level have already begun (Fig. 6.24). Between 1900 and the year 2000, the worldwide average sea level has already risen by 7.8 in. (20 cm). By the end of the twenty-first century, some scientists predict that sea levels worldwide may rise as much as another 2.6–6.6 ft. (0.8–2 m). This rise in sea levels is largely due to the rapid melting of glaciers worldwide, as well as the shrinkage of the ice-covered north and south polar regions. The rapid shrinkage of the ice cover at the North Pole now has the scientists

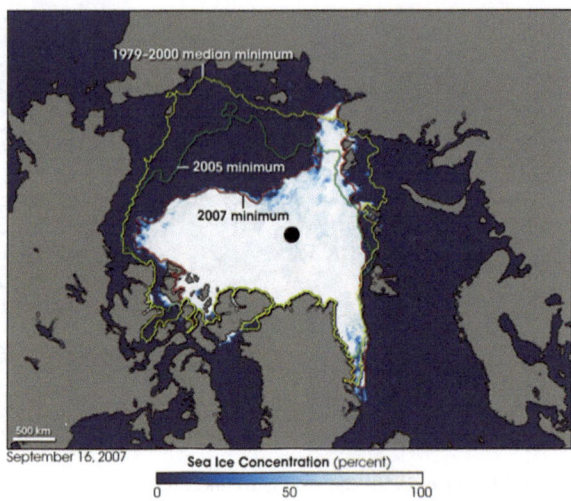

Fig. 6.25 NASA satellite photographs taken between 1979 and 2007 have revealed that the Earth's north pole (Artic) icecap is beginning to rapidly shrink in size due to global warming (image credit: NASA)

at NASA very concerned. Figure 6.25 shows the dramatic magnitude of this effect in space photographs taken between 1979 and 2007. Climatologists are now telling us that, with the present pace of global warming, the north polar oceans will start being completely ice free during the warmer summer months by the early 2050s The Greenland ice cap, which has been very stable in size for many eons, is also beginning to rapidly melt. If this dramatic loss of our ice and snow covers continues, the loss of coastal areas and low-lying areas worldwide, such as the American gulf coast and Holland, for example, will cause an economic, sociological, political, plus human nightmare scenario as entire populations will have to be displaced further inland. Global warming will, *and probably already has*, caused warmer ocean temperatures and more severe hurricanes and typhoons worldwide plus the recent sudden increase in extreme tornado activity in the United States midwest states.

Since 2005, the United States has experienced three of the worst hurricane events (hurricanes Katrina and Isaac, and "perfect storm" Sandy) ever recorded. When 100-year "super-storms" start hitting us every few years, it is time to start asking "why"? *Climatologists tell us that the hallmark of increased global warming is not increases in the mean annual numbers of bad weather events* (heat waves, cold waves, blizzards, ice storms, floods, droughts, tornadoes, hurricanes, etc.), *but in the severity levels of the events that do occur.*

Of course, not every expert accepts the possibility that mankind is headed toward a major catastrophe brought on by global warming. Still, the evidence continues to grow that this type of disaster may lie in our future unless we take immediate steps to cease our dependence on fossil fuels and switch to alternative clean energy sources. Many scientists strongly believe that the Earth's carbon cycle, which has for so many years allowed CO_2 levels to be controlled sufficiently to allow life to flourish, cannot be quickly turned on or off. If mankind stopped all usage of fossil fuels by 2050, it would still require several thousand years for airborne levels of

CO_2 to return to the levels that existed worldwide prior to 1900. The author sincerely believes that switching to clean energy alternatives is absolutely a "no-brainer". Fossil fuels are dirty and inefficient, dangerous to our health and, whether we like it or not, will be too inaccessible to be economically viable within one or two hundred years, anyway. The severity of future global warming effects may not be as bad as many present doomsday scenarios predict, but they definitely will not be good either.

However, one of the worst fears that a few scientists have expressed is that the continuing rise of CO_2 in the atmosphere might turn out to be the proverbial "straw that breaks the Camel's back" with respect to eventually triggering the worst case scenario of a runaway greenhouse effect (Ward 2007). A runaway greenhouse effect is what occurs when a planet's natural temperature regulating (thermostat) system gets completely swamped causing the surface temperature to begin rising in a totally uncontrollable fashion. The CO_2 greenhouse effect is not the only source of global warming that is occurring. The sun, itself, due to its natural evolution, is continuing over time to very slowly get hotter and hotter. And, as mentioned above, as the planet gets warmer and warmer, the resulting loss of ice and snow coverage on the surface causes more and more of the sun's light to be absorbed rather than being reflected back into space. Thus, the combined effects of the sun getting hotter, plus more sunlight being absorbed due to loss of reflective ice and snow, plus man's pumping more CO_2 into the air, may eventually turn out to be more than the Earth's thermostat system can handle. Our astronomers now believe a similar runaway greenhouse event was probably triggered on early Venus as a result of this planet's being slightly closer to the sun than is the Earth. Three or four billion years ago, when our sun was cooler, Venus might have been covered by liquid water which could have briefly allowed the emergence of primitive microbial life. Whether we like it or not, the natural tendency of all stars, including our sun, to gradually get hotter during their normal lifespans will eventually trigger a runaway greenhouse type phenomenon on our home planet. While this normal and unavoidable "act of nature" should not occur for another 5 billion or so years from now, a few scientists fear that, if mankind does not switch to clean energy, this doomsday scenario could happen much sooner.

Chapter 7
Some Final Thoughts from This "Amateur" Astronomer on Mankind's Imminent Discovery that We Are Not Alone in the Universe

My interest in astronomy began in the early 1950s when I was 10 years of age. Although I have been a devoted amateur astronomer all my life, I quickly learned that it would be necessary to find a real career that would "pay the mortgage". Fortunately, when I started college I also became fascinated by the subjects of psychology and the brain sciences which led me to obtain a Ph.D degree and pursue a professional career in the neurosciences. However, I never gave up my astronomy hobby. For 46 years now I have been a somewhat weird combination of an amateur astronomer and a professional neuroscientist. In more recent years, with the advent of computers, rocket science, and the exciting new field of astrobiology, I have discovered that my being an amateur and not a professional astronomer has provided me with what I believe may be a unique and important advantage in my personal understanding of astronomy and the space sciences. I believe that, as an amateur, I am more in tune with what the average non-scientists or hobbyists (i.e. people who spend rather than receive monies for pursuing their interest in astronomy) think and believe about the new field of astrobiology. When I use the term "mind boggling" in my writings, I am not trying to be a funny brain scientist, but rather am relating more closely to how many of the average citizens of our planet respond to this very fascinating new science of astrobiology.

Therefore, having been a weird (far out?) combination of both an amateur astronomer plus a professional neuroscientist for most of my life, I believe I may have developed some unique but possibly worthwhile ideas about what possible significance this exciting and totally fantastic idea that mankind may *not be alone in the universe* may have for our species. In the present chapter, I would like to put on my "amateur" astronomer hat and present some of my personal thoughts on not only how the discovery of intelligent extraterrestrial life forms (biological or *possibly* even non-biological mechanical systems) may influence mankind's emotional or psychological states, but also tackle a few topics that I believe are the source of the mind blowing response that many of us amateurs experience (and a few professionals deny) when first encountering some of the more unusual, puzzling, or even "magical" aspects of this literally "other world" science. I am hoping

© Springer International Publishing Switzerland 2015
J.L. Cranford, *Astrobiological Neurosystems*, Astronomers' Universe,
DOI 10.1007/978-3-319-10419-5_7

that my thoughts or ideas on some of these topics will assist my non-science readers in relating to the larger and more profound effect that all this talk by our scientists about the evolution of intelligent ETs may have for our beliefs on why we and the universe are here.

Without a doubt, the single factor that is the greatest source of confusion for most of us (non-scientists as well as scientists) when it comes to understanding how the universe works is that totally bizarre concept that many scientists refer to as *deep time* (Darling 2013). The universe is definitely not young in any sense of the word! Many very sincere and religious people believe it only took 7 days for God or some higher power to create the entire universe, including Earth, the stars, and everything else that exists. Up until just the last few hundred years, most people, including many of our best scientists even believed that Earth itself could not be more than just a few thousand years old. However, our scientists now tell us that the entire universe, including everything we can see with our largest telescopes, may be as much as **13.7 billion** years old, and some scientists have recently started speculating that it may even be older, with a few serious scientists even going so far as to suggest the possibility that it might be infinite with respect to both age and size, or that there could be more than one universe (i.e., multiple universes, parallel universes), or even other universes based on physical dimensions that are different from the three traditional dimensions plus time from which our own universe is constructed (Kaku 2006). And, to make things even more bizarre, *a few scientists have even dared suggest the possibly that some of these other universes may host totally different laws of physics and chemistry from our own universe* (Gribbins 2009; Greene 2011; Vilenkin 2006).

Since all of us live for an extremely short period of time on a small planet in an incredibly huge universe that is unbelievably old, it is no wonder that none of us, including our smartest scientists, have any rational concept of what this thing we call "*time*" is (Whitrow et al. 2004). Our psychologists tell us that, while most of us can make sense of what it means to be young or old in terms of our own lifespans (age), virtually all of us have difficulty relating to any historical or future events that occurred more than just a few hundred years before we were born or may happen following our death. Therefore, when our scientists inform us that the huge hostile universe we live in was itself born in some kind of "Big Bang Event" close to 13.7 billion years ago, and that our own tiny fragile home planet Earth was not created until much later at about 4.5 billion years ago, it comes as no surprise that many of us might become somewhat depressed or even humbled when those same scientists also tell us that if we are very healthy and even luckier, we might be able to live to the ripe old age of 70 or 80 years, or even older. The average citizen's idea of "time" and the scientist's concept of "deep time" are definitely quite different!

So, how confusing would it be if some scientist (Schopf 2002) presented us with an analogy that compared the age of the Earth to a single 24-h day? For purposes of such an "analogy" (Fig. 7.1), instead of the Earth being 4.5+ billion years old, we would have to assume that the Earth's birth occurred early one morning at the "bewitching hour" of 12 a.m. And that first living ancient single-celled ancestor of ours (that lived 3.8 billion years ago), would not evolve and start doing its

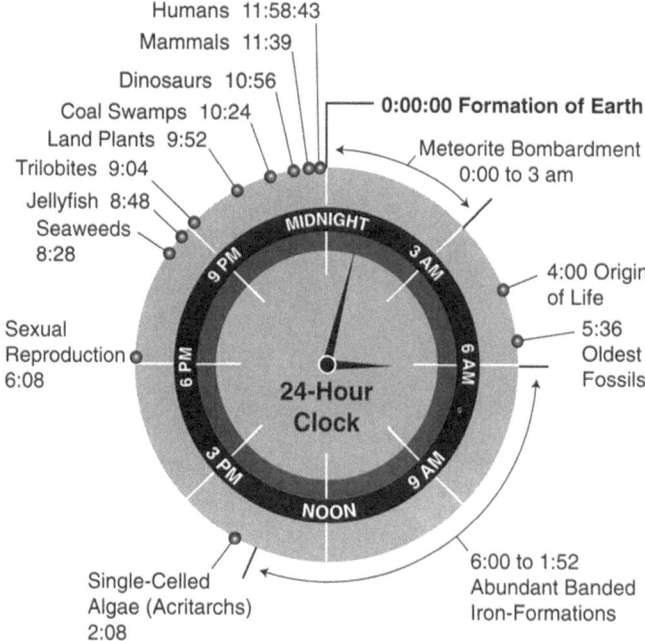

Fig. 7.1 Shows one life scientist's depiction of how the creation of Earth and the evolution of life would appear if contrasted with that of one normal 24-h day rather than the 4.5+ billion year process that was actually involved (image credit: Dr. William Schopf)

swimming thing in our hot oceans until 4 h later at about 4 a.m. After that, things would not change very much (life would continue being single-celled and microscopic, i.e., too small to be seen with the naked eye) until the following evening at about 7 p.m. when the first macroscopic (visible to the naked eye) multi-cellular life forms would make their debut on Earth. After that, life would seem to virtually "explode" in terms of size, complexity, and diversity with all sorts of new multi-celled organisms evolving right and left until, just before 11 p.m. (i.e., a tad more than 1 h before the end of our 24 h day), the dinosaurs would make their debut. And then, with slightly less than a minute and a half before midnight, the first cave dwelling humans would arrive, and with less *than 1 s left before the stroke of midnight*, human civilization would get started.

Finally, before concluding our brief discussion of deep time, I would like to suggest that the reader look at Fig. 7.2, which presents a recently updated and expanded version of the classic Cosmic Calendar analogy that Carl Sagan presented in his excellent 1977 book (Sagan 1986). In contrast to Dr. William Schopf's (Schopf 2002) use of a 1-day 24-h model for his deep time analogy, Dr. Sagan's Cosmic Calendar uses a 12-month calendar analogy that presents a much broader comparison of the astronomical events that occurred from the birth of the universe or the Big Bang event itself right up to the creation of the Earth and the evolution of life and the rise of human civilization. While Dr. Schopf's 24-h model is simpler

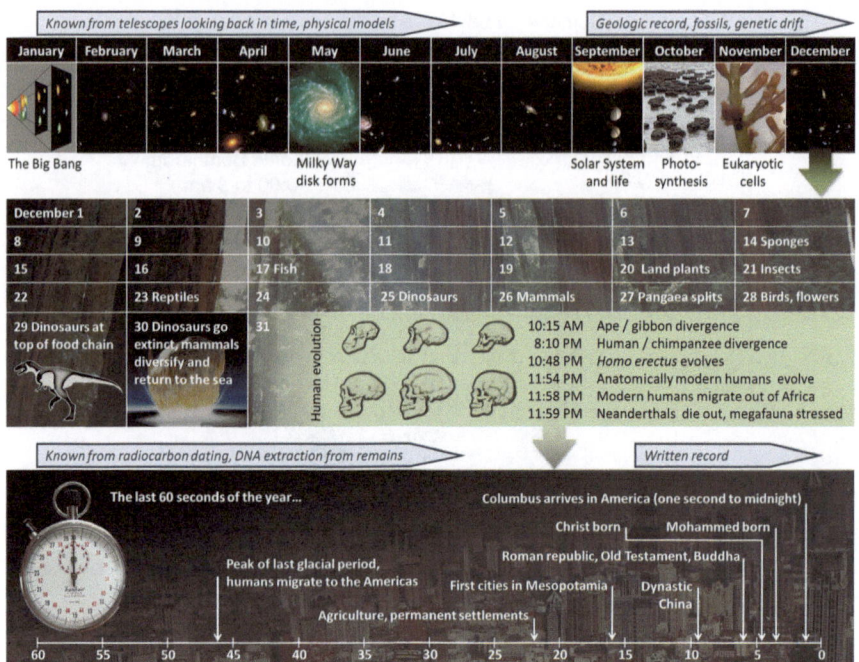

Fig. 7.2 Shows a recent updated version of Carl Sagan's classic *"Cosmic Calendar"* analogy in which he contrasts the astronomical/biological events that occurred following the Big Bang Inflation Event right up to Christopher Columbus' discovery of America (image credits: The image is original work by Eric Fisk, with public domain Images composited from Wikipedia Commons, NASA, and other listed government sites)

and easier to quickly grasp, Dr. Sagan's analogy will provide interested readers with more information related to the complexity of this deep time thing, although it may boggle your brain even more (as it did mine) than did Dr. Schopf's model.

In addition to our universe being "unbelievably old", another thing that our astronomers and other scientists have definitely laid on man's fragile human psyche in the last century or two is the fact that our universe is, in a very real sense, also extremely hostile in terms of allowing either non-living or living matter to remain totally unchanged for long periods of time.[1] Nothing remains the same as "it or they are right now" i.e., in a *stable status quo* with the surrounding environment, forever. Rocky planets are occasionally (albeit every few million Earth years!) slammed

---

[1] Although our scientists and science teachers have been telling us for many years that we live in an extremely hostile universe that could easily destroy all life on our planet at any time, the average adult citizens of our world would not freak out in any way to this idea unless they saw it happening before their own eyes. It appears that science's concept of deep time has totally made most of us immune to any fears of imminent destruction from the skies or from beneath the Earth. People still live quite comfortably and fearlessly at the base of dangerous volcanoes or in "severe" tsunami or earthquake zones.

into by asteroid or comet impacts, ravaged by intense super volcanoes, zapped by deadly radiation from nearby exploding stars, or eaten by black holes left behind by those same dying stars, or even rendered uninhabitable by their intelligent but misguided residents. And stars can only maintain a stable lifestyle or what astronomers call a "main sequence" phase for a limited time (once again, involving many millions or billions of Earth years!) before running out of the nuclear fuel that they need to keep on shining and begin a slow dying process ending in a small dense piece of rock called a "white dwarf" in the case of smaller stars like our sun, or exploding in a horrific bright and violent explosion (called super- or hypernovae) in the case of larger stars. And, of course, we all know that plants and animals can only stay viable and healthy as long as their internal biological aging clocks allow or until the next earthquake, ice age episode, runaway greenhouse event, or some manmade disaster or catastrophe occurs that renders their world uninhabitable.

Many scientists, including me, also believe that, at least on our own home planet, nature has only two ways of allowing living creatures to deal with their hostile environments in terms of being able to successfully survive and reproduce. The first method is biological evolution in which random changes or alterations (mutations) in the biological (genetic) makeup of individual living organisms (or whatever ET's equivalent process may involve) produce physiological and morphological changes that allow the offspring of such organisms to successfully compete in the reproduction/survival game. Nevertheless, while the evolutionary mode of long term adjustment to changing environments is, at least on our planet, both unbelievably slow and "inefficient"(?), if given, of course, enough "*deep*" time (thousands, millions, or more likely, billions of Earth years) it can and does seem to work quite well, at least from the viewpoint of newly emerging species. All the reader needs to do to verify how successful this evolutionary mode of adjustment to hostile environments can be is to look around you anywhere you want—life (lots of it) is everywhere on our planet, in every color, shape, and size, etc. And to further emphasize this point, our scientists tell us that the estimated 10–100 million species of different animals and plants that are alive on Earth today represent only about 1–5 % of the total number of species that have lived on Earth since the first living creatures formed circa 4 billion years ago. Thus, the history of life on our own home planet has been, over the ages, an absolute "whirlwind of the comings and goings" of incredible varieties of different life forms in direct response to a multitude of natural biological evolutionary changes and occasional catastrophic geologic or cosmic events![2]

And, of course, the second mode of adjustment to hostile environments that nature provides for us earthlings involves the development by both animals and plants of specialized biological systems that allow them to quickly respond to more

---

[2] Although life is still extremely plentiful everywhere on our planet, most scientists believe that the evolution of life has now peaked and is on the decline, and will disappear entirely in a few billion years as our sun continues its normal "warm-up" thing. While all planets probably undergo similar longterm changes in response to this normal evolution of their home stars, man's incessant pollution of his environment is likely speeding this process up quite rapidly.

sudden or frequent changes in their environment to avoid destruction. While living matter can actively respond or adjust to its environment in order to survive and reproduce, non-living matter can do virtually nothing but passively "weather the storm", so-to-speak, or convert from one physical state to another (e.g., water can freeze into ice, rocks can be eroded into sand, and even inhabited planets are vulnerable to being "eaten alive" by wandering black holes). At least on our planet, the range of responses that different living species (both plants and animals) are capable of making to changes in their world are amazing. In describing this unique feature of living matter, many scientists use the concept of "biological intelligence" to describe this truly profound natural phenomenon. While psychologists, psychiatrists, theologians, educators, etc., typically use the term intelligence to only refer to the mental functions (i.e., thinking, learning, remembering, etc.) of humans and certain other advanced animal species (e.g., apes, porpoises, elephants), the author, in spite of being a psychologist by training, has expanded his concept of "biological intelligence" to include the ability of all living organisms (be they carbon-based plants and animals, or ETs that are carbon-based, silicon-based, boron-based, or whatever) to do much more than just interact with their environment via mental or psychological (behavioral) processes to adjust to changes in their worlds. My involvement with astronomy and, more recently, astrobiology, therefore, makes me include, as part of my own personal definition of biological intelligence, not only the ability of the brains or nervous systems of animals to ensure their owner's survival, but also the equivalent non-neural biological systems of plants (plus, perhaps, many ETs) that allow these life forms to also interact with their surroundings and make appropriate physiological adjustments to maintain their owner's status quo in the world (i.e., keep the environment from harming or destroying them).

Therefore, my personal concept of biological intelligence also includes vegetative (more plant-like or "non-mental") adjustments of organisms to environmental changes. When it rains or storms, animals seek shelter, plants turn their leaves toward sunlight to obtain energy or fold up their flower petals to conserve moisture when the climate gets too dry, and seeds (or bears) go into suspended animation or hibernation states, etc. And, of course, so-called more advanced or "intelligent" life forms such as humans can go to the doctor and get a penicillin shot when they have an infection, or invent machines (artificial intelligence) that can give them relief from the mundane chore of always having to be personally involved in thinking or problem solving tasks. Thus, my "hybrid" science background also makes me believe humans, plants, and ETs all develop special biological systems that allow them to "quickly" do much more than think, remember, recognize, or problem solve, etc. They also exhibit many other types of responses which are also complex but, for obvious lack of any better words, are sometimes loosely referred to as "mindless reflexes" (or pre-wired and inherited instinctive adjustments) to keep nature's "grim reaper" at bay. Of course, the evolutionary development of complex neurological or biological systems for the primary purpose of allowing living organisms to adjust and survive in the world has, in the case of humans at least, provided us with a very important secondary benefit. Our brains also allow us to improve our lives (e.g., invent music and poetry to entertain us, medicines to keep us healthy, and air conditioning to keep us comfortable) as well as unfortunately sometimes harm us (e.g., polluting and destroying our environment, inventing nuclear weapons to protect us from imaginary foes, etc.).

Some readers, and no doubt many scientists, might take issue with my describing some of the simpler "behavioral" adjustments of some plants, such as turning their leaves toward the sun to collect more light, as being a form of biological "intelligence". I do believe, however, that it is entirely possible that ET might be able to evolve on other worlds in some kind of "dual format" similar to our own plants and animals. Who is to say that in a galaxy far far away, our astrobiologists may not someday discover some truly bizarre ETs in which plant-like creatures are actually "smarter" than their animal-like cohorts! Some respected Earth-bound plant scientists or botanists (Baluska et al. 2006; Muller et al. 2006; Trewavas 2006; Volkov 2012) have, in recent years, started finding that some plants, in spite of not having a brain or nervous system, appear to be capable of responding to their environments or other plants in fashions remarkably similar to what animals do. Many species of plants appear to be able to manufacture and use special chemicals to engage in fighting predators, maximizing food opportunities, and even competing with other plants. This research, while still very controversial, is expanding and is now being accepted as possible by increasing numbers of scientists.

As the reader has probably surmised, the author also believes, based purely on informed speculation but presently no solid experimental evidence that life, including intelligent life, is common (although, paradoxically, probably mathematically rare) throughout our unbelievably vast and hostile universe and will likely turn out to be very diverse in terms of its physiological and chemical constitutions or makeup, and even its behavioral manifestations. And, I also believe that our scientists will, before the end of the new twenty-first century, also confirm the existence of past and/or present life in more than one or two locations other than Earth within our own solar system. These other life forms which live (or lived) in our own backyard will also likely be determined to be truly aliens, i.e., independent geneses (origins) of life that are different from that which evolved on Earth and, although it is possible, these other life forms will probably not be the result of some primitive microbial life forms relocating from one planet (or moon) to another by hitching rides inside meteors (i.e., panspermia). And, less the reader or others think I am totally arrogant in making these predictions, let me make it perfectly clear that I am in no way the first amateur or professional scientist to take this leap. With the rise of computers and the space age, and especially the recent launching of the NASA Kepler space telescope, my voice is but one of many that are now shouting from the rooftops of laboratories (or even observatories) all over the world.[3]

---

[3] However, as a trained scientist, I must blame my overly enthusiastic response in parts of this book as being the result of my "amateur" astronomer status and not my professional training in science. Scientists must always be conservative and never declare any research evidence as being any more than "highly suggestive" in supporting some conclusion or hypothesis as to how something works. As an amateur, I feel somewhat more free to join the ranks of our science fiction cohorts and

The surge of indirect but solid scientific evidence for the mind blowing possibility that life may be quite common in our vast universe (Darling 2001; Dartnell 2007) literally exploded onto the world stage in the 1970s and is continuing to develop, if such a thing is possible, at an even faster pace today. For the majority of the human citizens of planet Earth, the confirmation by our scientists that ET actually exists somewhere out there in space will, no doubt, definitely be an "eye opener". Of course, the magnitude (emotional intensity) of mankind's response will vary widely depending on the specific manner in which this news is released to the public. This historical event might arise in the form of a simple announcement from NASA/ESA or a press conference by the President of the United States or the Prime Minister of England that a "twin-Earth" type planet has been identified that appears to possibly host all of the prerequisite conditions needed to support our form of carbon-based life. A much more extreme reality check might come in the form of something akin to a Steven Spielberg "close encounters of the third kind" scenario involving ETs actually landing on the White House lawn and stepping out of their space ships to greet us (or destroy us). In between these two extremes are a large number of alternative possibilities that will trigger a wide range of responses from our citizens that will most likely differ depending on each individual's personal, educational, and professional background. The announcement that single-cell organisms similar to our bacteria exist on Mars would definitely cause our life and space scientists to go bonkers, but many non-scientists would probably do little more than smile and shrug their shoulders. The announcement, however, that a far distant Earth-like planet actually hosts intelligent multi-cellular creatures with advanced technologies would most likely change many of those smiles and shrugs into more advanced stages of emotional excitement, whether pleasant or otherwise (e.g., fear, panic). The revelation that we actually are not alone in the universe would undoubtedly be so profound that no psychologist, theologian, scientist, or anyone else alive today would be able to predict how mankind will respond to this news (Davies 1995; Jakosky 2006). We will simply have to wait and see what happens, and hope that the response will, for the most part, be positive and constructive rather than detrimental to our human society.

Given mankind's long and quite pervasive history of being predatory and emotional creatures, would our beliefs in a God, Higher Power, or some kind of Creative Force in the universe change? For many of us, most definitely! However, once again, the kind, direction, intensity, etc., of such religious/spiritual transformations would vary dramatically from person to person. Some individuals would suddenly become religious for the first time, or become atheists (or agnostics), or see their religious beliefs dramatically intensify. Others would simply see the confirmation of the existence of ET as strong additional evidence that the universe may be at least partly under the control of some kind of "higher

---

express my enthusiasm in a less conservative and professional manner, *as long as I am careful to make sure my listeners or readers know the difference.*

power" or non-physical (spiritual) force(s). Many people would probably have their religious "cages" rattled to some extent!

Would the "dark side" of mankind's nature, i.e., our long standing historical "compulsion" that makes otherwise friendly and sane people go to war every few years with their neighbors and try to kill each other for any of a wide range of both "good" and "bad" excuses, including patriotism, religious or cultural differences of opinion, removal of evil tyrants, acquiring more land and natural resources, whatever, go away if we suddenly realized that we are not alone in the universe? In my opinion, two of the greatest concepts that can at different times be either "good" or "bad" for mankind are the ideas of patriotism and national borders. In my opinion, the "*Doctors without Borders*" organization of physicians is one of the few human groups that are doing it right. If mankind suddenly discovers that he is but one of a multitude of intelligent life forms that are living on very fragile planets in the universe, would we rethink our concepts of sovereign nations, patriotism, or national borders? From a purely logical (or even religious/spiritual) point of view perhaps all we should be "worshiping" is life itself and not all the other incidental physical trappings that come attached to it, i.e., money, personal property, national territories, etc. Just maybe, life itself is the true higher power or creative force in our vast universe, and not all the material or physical stuff that surrounds us.

Finally, because of my unusual background as an astronomer/brain scientist, I would now like to add a few more of my own personal thoughts that might possibly be important to the theme of the present book. As the reader is by now well aware, my adult vocation (Ph.D) was in the areas of psychology and the brain sciences. For close to 40 years I struggled to learn as much as I could about these very complex subjects. Although we know a great deal about these topics, any professional (and especially me) will tell you that there is much more that we do not yet know. Even with all the amazing technological breakthroughs in the sciences that are needed to understand how the brain works, what we actually know (or think we know) is less than a tiny fraction of what is yet to be discovered. This incredible reality check is also applicable to the author's other passions including the fields of astronomy and the even newer field of astrobiology.

Unfortunately, the biggest handicap to our understanding how the universe and life works is that all of us were born on the same extremely tiny planet and learned to both communicate and think with whatever language system we were exposed to at an early age. In a very real sense, all of us earthlings are virtual "slaves" to our native language systems which, at least in our early formative years, did not provide us with much in the way of words or other linguistic concepts that allowed us to acquire any understanding of astronomy and the life sciences. Our first exposure to the advanced version of our language system that did contain vocabulary items, concepts, or words which allowed us to begin thinking, writing, and speaking about these topics probably did not occur until later in elementary school or in high school. For those of us that got hooked on these subject areas, our brains suddenly started being inundated by numerous strange (and long-winded) words, concepts, ideas, and thoughts in college and then totally overwhelmed later in our graduate school years and professional careers.

Since language is a means by which our brains actively perceive and develop some kind of semi-permanent internal representations or "models" (physical, chemical/molecular, or electrical "memories") of events or happenings that we personally experience in the real world and then allow us to transfer these "models" to others that may have experienced similar events, we encounter few communication problems. Nevertheless, when our astronomers (and brain scientists) started encountering strange new phenomena/events in their laboratories and observatories, they found they had no words with which to transfer these "experiences" to their non-science friends, students, and even colleagues. It was as if they were now speaking a foreign language which, of course, they were. They had to start making up or fabricating completely new words and concepts. In a very real (and frustrating) sense it is extremely difficult for anyone to verbally describe something that they themselves may have experienced but their listeners (readers) have not experienced. Scientists, therefore, being enslaved to their own native languages, are extremely handicapped whenever they attempt to give lectures or write books in their specialty areas.

In the present book, I have attempted to transfer to the readers my own complex (and, unfortunately, possibly confusing internal neural models) of what our best scientists now tell us is about to become mankind's single greatest and most profound reality check—the discovery that we are not the only living creatures in the universe! Up until now, all of this has been purely the stuff of science fiction, although tons of fun for those of us who enjoy having our imaginations expanded and tweaked. However, for all of us, the time is rapidly approaching when we, as humans, will need to stop acting like predatory idiots and get our act together—*IF WE WANT TO SURVIVE LONG ENOUGH TO JOIN THE UNIVERSAL COMMUNITY OF LIFE*! I personally am hoping that this brave new world of ETs will be life friendly and not, like us, predatory. If not, it might be best for us to keep quiet and hope that "they" do not discover our existence and choose to visit us for something other than peaceful purposes.

Even if we were to discover that we actually are alone in the universe, I would hope that we humans would still be willing and eager to make any and all changes to our way of life that are necessary to at least ensure our long term survival or, perhaps, when our sun becomes too hot for our comfort, expand our society to cooler parts of our solar system, or to other twin Earths elsewhere in the universe. Even if our fellow humans (plus our non-human relatives) were the only company we have in the universe, mankind should be eager to do whatever is necessary to keep all of us happy and healthy for as long as possible. Crimes and wars should be eliminated, everywhere. National borders can remain in place, but gates and armed border guards need to be eliminated. As a human being and most definitely as an educator, I have long believed that quite possibly the greatest and most rewarding asset that mankind possesses is our widespread and extremely rewarding historical, cultural, religious, and racial diversity. Independent sovereign nations are good things as long as they are freely accessible to everyone on the planet, and their citizens have no concepts of "outsiders" or "foreigners". Friendly competition between individuals, professional or social groups, or countries is good for our souls, minds, and pocket books. Even debate or differences of opinion can be allowed as long as they are positive. Such activities definitely can be stimulating and productive, plus frequently entertaining.

However, as may be already apparent to the reader by the tone of many of my statements in the present book, I must confess that I am unfortunately not alone among many other concerned citizens of our world in my gut level fear (hopefully unfounded) that both the best and worst things that could have happened to our small and fragile planet is the emergence of mankind. As Peter Ward (2009) in his new book entitled *"The Medea Hypothesis: Is Life on Earth Ultimately Self-Destructive"* has so eloquently and powerfully suggested, almost all life forms, because of their inherent natures, compete with each other and try to eliminate each other and, in doing so, end up damaging or even destroying their environments. In the author's opinion (which is shared by many others), mankind, because of both our emotional/predatory as well as intelligent natures, needs to be placed at the top of that self-destructive list. It does seem that the evolution of intelligence and the tendency for self-destruction, at least in our species, is highly correlated. Hopefully, this correlation does not reflect a direct cause and effect relationship (ergo, some kind of hardwired genetic based neural entity) but is due to other factors that mankind might be able to control or eliminate, if we collectively choose to do so!

And now, for my final personal thought (or wish). I strongly and sincerely believe that, whether we are alone in the universe or not, we absolutely must rid ourselves of our predatory tendencies (whether hardwired or, hopefully, not) and begin doing whatever is necessary to change mankind's trek towards possible self-destruction? I believe a good place to start would be to eliminate mankind's inherent "tribal warfare" mentality (including crime and terrorism), and our impending global warming crisis. Perhaps channeling the world's military/defense budgets into monies to educate our children, eliminate poverty and hunger, promote medical research to make people healthier, and support the fine arts and sciences to make people happier and more informed, would make for a great beginning. I personally hope that the discovery that we have potential friends and colleagues out there in space might provide us a new and powerful incentive to make this long-needed but now absolutely critical transformation. If we do not, we will probably perish long before the normal heating up of our sun ends mankind's short reign on our planet.

References and Further Suggested Readings

Aguilar, D.A.: Alien Worlds: Your Guide to Extraterrestrial Life. National Geographic for Kids, Washington, DC (2013)

Al-Chalabi, A., Turner, M.R., Delamont, R.S.: The Brain: A Beginner's Guide. Oneworld, Oxford, UK (2006)

Amthor, F.: Neuroscience for Dummies. Wiley, Mississauga, ON (2012)

Asimov, I.: The Human Brain: Its Capacities and Functions, Revised and Expanded Edition. Penguin, New York (1994)

Ballesteros, F.J.: ET Talk: How Will We Communicate with Intelligent Life on Other Worlds? Springer, New York (2010)

Baluska, F., Mancuso, S., Volkmann, D.: Communication in Plants: Neurological Aspects of Plant Life. Springer, Heidelberg (2006)

Baross, J.: Evolution: a defining feature of life. In: Sullivan, W.T., Baross, J. (eds.) Planets and Life: The Emerging Science of Astrobiology, pp. 213–231. Cambridge University Press, New York (2007)

Beaumont, J.G.: Introduction to Neuropsychology. Guilford, New York (2008)

Benner, S., Davies, P.: Towards a theory of life. In: Impey, C., Lunine, J., Funes, J. (eds.) Frontiers in Astrobiology, pp. 25–47. Cambridge University Press, New York (2012)

Bennett, J.: Beyond UFOs: The Search for Extraterrestrial Life and Its Astonishing Implications for Our Future. Princeton University Press, Princeton, NJ (2008)

Bennett, J., Shostak, S.: Life in the Universe, 3rd edn. Addison Wesley, San Francisco, CA (2011)

Bennett, J., Shostak, S., Jakosky, B.: Life in the Universe. Addison Wesley, San Francisco, CA (2003)

Bortz, F.: Astrobiology: Cool Science. Lerner, Minneapolis, MN (2008)

Bremner, J.D.: Brain Imaging Handbook. W.W. Norton, New York (2005)

Brooks, R., Fritz, S., et al.: Invited Panelists for Scientific American, Understanding Artificial Intelligence. Scientific American, New York (2002)

Brownlee, D.E., Kress, M.: Formation of earth-like habitable planets. In: Sullivan, W.T., Baross, J., (eds.) Planets and Life: The Emerging Science of Astrobiology, pp. 69–90. Cambridge University Press, New York (2007)

Casoli, F., Encrenaz, T.: The New Worlds: Extrasolar Planets. Springer-Praxis, New York (2007)

Chaisson, E., McMillan, S.: Astronomy Today, 3rd edn. Prentice-Hall, Upper Saddle River, NJ (2000). More recent editions available

Charlesworth, B., Charlesworth, D.: Evolution: A Very Short Introduction. Oxford University Press, Oxford, UK (2003)

© Springer International Publishing Switzerland 2015
J.L. Cranford, *Astrobiological Neurosystems*, Astronomers' Universe,
DOI 10.1007/978-3-319-10419-5

Chela-Flores, J.: The New Science of Astrobiology: Cellular Origins and Life in Extreme Habitats to Evolution of Intelligent Behavior in the Universe. Kluwer, Boston, MA (2001)

Chela-Flores, J.: Habitability on Kepler worlds: are moons relevant? In: de Vera, J.-P., Sechback, J. (eds.) Habitability of Other Planets and Satellites (Cellular Origins, Life in Extreme Habitats, and Astrobiology), pp. 149–366. Springer, New York (2013)

Chyba, C., Phillips, C.: Europa: potentially habitable world. In: Sullivan, W.T., Baross, J. (eds.) Planets and Life: The Emerging Science of Astrobiology, pp. 388–423. Cambridge Universisty Press, New York (2007)

Cohen, J., Stewart, I.: What Does a Martian Look Like? The Science of Extraterrestrial Life. Wiley, Hoboken, NJ (2002)

Copley, S., Summons, R.: Terran metabolism: The first billion years. In: Impey, C. (ed.) Frontiers of Astrobiology, pp 48–72. (2012)

Courtillot, V., McClinton, J.: Evolutionary Catastrophes: The Science of Mass Extinctions. Cambridge University Press, New York (1999)

Coustenis, A., Blanc, M.: Large habitable moons: Titan and Europa. In: Impey, C., Lunine, J., Funes, J. (eds.) Frontiers in Astrobiology, pp. 175–200. Cambridge University Press, New York (2012)

Coustenis, A., Encrenac, T.: Life Beyond Earth: The Search for Habitable Worlds in the Universe. Cambridge University Press, New York (2013)

Cranford, J.L.: From Dying Stars to the Birth of Life: The New Science of Astrobiology and the Search for Life in the Universe. Nottingham University Press, Nottingham, UK (2011)

Darling, D.: Life Everywhere. Basic, New York (2001)

Darling, D.: Deep Time: The Journey of a Particle from the Moment of Creation to the Death of the Universe and Beyond. First Edition Design, Sarasota, FL (2013)

Dartnell, L.: Life in the Universe: A Beginner's Guide. Oneworld, Oxford, UK (2007)

Davies, P.: Are We Alone? Philosophical Implications of the Discovery of Extraterrestrial Life. Basic, New York (1995)

Davies, P.: The Eerie Silence: Reviewing Our Search for Alien Intelligence. First Mariner Books Edition, New York (2011)

Delsemme, A.: Our Cosmic Origins: From the Big Bang to the Emergence of Life and Intelligence. Cambridge University Press, Cambridge (1998)

Dreamer, D.: First Life: Discovering the Connections Between Stars, Cells, and How Life Began. University of California Press, Berkeley, CA (2011)

Dubin, M.: How the Brain Works. Blackwell, Malden, MA (2002)

de Duve, C.: Vital Dust: The Origin and Evolution of Life on Earth. Basic, New York (1995)

de Duve, C.: Life Evolving. Oxford University Press, Oxford, UK (2002)

Ehrenfreund, P., et al. (eds.): Astrobiology: Future Perspectives (Astrobiology and Space Science Library). Kluwer, Norwell, MA (2004)

Ekers, R.D., Kent Cullers, D., Billingham, J., Scheffer, L.K. (eds.): SETI 2020: A Roadmap for the Search for Extraterrestrial Intelligence. SETI Press, Mountain View, CA (2003)

Erwin, D.H.: Extinction: How Life on Earth Nearly Ended 250 Million Years Ago. Princeton University Press, Princeton, NJ (2006)

Fastovsky, D., Weishampel, D.: Dinosaurs: A Concise Natural History, 2nd edn. Cambridge University Press, Cambridge, UK (2012)

Futuyma, D.J.: Evolution. Sinauer, Sunderland, MA (2005)

Gardner, J.: Biocosm: The New Scientific Theory of Evolution; Intelligent Life Is the Architect of the Universe. Inner Ocean, Maui, HI (2003)

Gardner, J.: The Intelligent Universe: AI, ET, and the Emerging Mind of the Cosmos. New Page, Franklin Lakes, NJ (2007)

Gargaud, M., Martin, H., Lopez-Garcia, P., Montmerle, T., Pascal, R., Dunlop, S.: Young Sun, Early Earth, and the Origins of Life: Lessons for Astrobiology. Springer, Heidelberg (2009)

Gilmour, I., Sephton, M.A.: An Introduction to Astrobiology. The Open University Press, Cambridge, UK (2003)

Gould, S.J.: The Book of Life. W.W. Norton, New York (2001)

Greene, B.: The Hidden Reality: Parallel Universes and the Deep Laws of the Cosmos. Alfred A. Knoff, New York (2011)

Grenfell, J.I., et al.: Exoplanets: criteria for their habitability and possible biospheres. In: de Vera, J.-P., Sechback, J. (eds.), Habitability of Other Planets and Satellites (Cellular Origins, Life in Extreme Habitats, and Astrobiology), pp 13–30. Springer, New York (2013)

Gribbins, J.: Multiverse: Parallel Worlds, Hidden Dimensions, and the Ultimate Quest for the Frontiers of Reality. Wiley, Hoboken, NJ (2009)

Haines, T., Chambers, P.: The Complete Guide to Prehistoric Life. Firefly, Richmond Hill, ON (2005)

Haines, S., Sahtouris, E., Swimme, B.: A Walk Through Time: From Stardust to Us. Wiley, New York (1998)

Hallam, T.: Catastrophes and Lesser Calamities: The Causes of Mass Extinctions. Oxford University Press, Oxford, UK (2005)

Hamilton, W.B.: Plate Tectonics and Man. Report from USGS Annual Report, Fiscal Year. U.S. Geological Survey, Reston, VA (1976)

Hazen, R.M.: The Story of Earth: The First 4.5 Billion Years, from Stardust to a Living Planet. Penguin, New York (2013)

Holley, J.W.: How Likely Is Extraterrestrial Life? Springer, New York (2012)

Impey, C., Lunine, J., Funes, J.: Frontiers of Astrobiology. Cambridge University Press, Cambridge, UK (2012)

Irwin, P.G.T.: Detection methods and properties of known exoplanets. In: Mason, J. (ed.) Exoplanets: Detection, Formation, Properties, Habitability, pp. 1–16. Springer, New York (2008)

Irwin, L.N., Schulze-Manuch, D.: Cosmic Biology: How Life Could Evolve on Other Worlds. Springer Praxis, New York (2010)

Jakosky, B.: The Search for Life on Other Planets. Cambridge University Press, Cambridge, UK (1998)

Jakosky, B.: Science, Society, and the Search for Life in the Universe. University of Arizona Press, Tucson, AZ (2006)

Jakosky, B., Westall, F., Brack, A.: Potentiall habitable worlds (Mars). In. Sullivan, W.T., Baross, J. (eds.) Planets and Life: The Emerging Science of Astrobiology, pp 357–384. Cambridge University Press, New York (2007)

Jastrow, R., Rampino, M.: Origins of Life in the Universe. Cambridge University Press, New York (2008)

Jayawardhana, R.: Strange New Worlds: The Search for Other Planets and Life Beyond Our Universe. Princeton University Press, Princeton, NJ (2011)

Jerome, K.B.: Understanding the Brain. National Geographic Books, Margate, FL (2003)

Jones, B.W.: Life in the Solar System and Beyond. Springer-Praxis, Chichester, UK (2004)

Jones, B.W.: The Search for Life Continued: Planets Around Other Stars. Springer-Praxis Books in Popular Astronomy, Berlin (2011)

Kaku, M.: Parallel Worlds: A Journey Through Creation, Higher Dimensions, and the Future of the Cosmos. Anchor, New York (2006)

Kalat, J.: Biological Psychology. Wadsworth Centage, Belmont CA (2008)

Kasting, J.F.: How to Find a Habitable Planet (Science Essentials). Princeton University Press, Princeton, NJ (2010)

Kaufman, M.: First Contact: Scientific Breakthroughs in the Hunt for Life Beyond Earth. Simon & Schuster, New York (2011)

Kean, L.: UFOs: Generals, Pilots, and Government Officials Go on the Record. Harmony, New York (2010)

Knoll, A.H.: Life on a Young Planet: The First Three Billion Years of Evolution on Earth. Princeton University Press, Princeton, NJ (2003)

Knoll, A.H., Canfield, D.E., Konhauser, K.O. (eds.): Fundamentals of Geobiology. Wiley, Oxford, UK (2012)

Krukonis, G., Barr, T.: Evolution for Dummies. Wiley, Hoboken, NJ (2008)

Kurzweil, R.: The Age of Spiritual Machines: When Computers Exceed Human Intelligence. Penguin, New York (2000)

Kurzweil, R.: The Singularity Is Near: When Humans Transcend Biology. Penguin, New York (2006)

Kwok, S.: Organic Molecules in the Universe. Wiley, New York (2012)

Kwok, S.: Stardust: The Cosmic Seeds of Life. Springer, Heidelberg (2013)

Lequeux, J.: Birth Evolution and Death of Stars. Sinauer, Hackensack, NJ (2013)

Liebman, M.: Neuroanatomy: Made Easy and Understandable. Aspen, Rockville, MD (1986)

Louis, C., Minneli, D.: Review of known exoplanets. In: Impey, C., Lunine, J., Funes, J. (eds.) Frontiers in Astrobiology, pp 250–268. Cambridge University Press, New York (2012)

Lunine, J.I.: Earth: Evolution of a Habitable World. Cambridge University Press, Cambridge, UK (1999)

Lunine, J.I.: Astrobiology: A Multidisciplinary Approach. Pearson, Addison Weslely, New York (2004)

Lunine, J., Butler, P.: Titan: potentially habitable world. In: Sullivan, W.T., Baross, J. (eds.) Planets and Life: The Emerging Science of Astrobiology, pp. 424–442. Cambridge University Press, New York (2007)

MacLean, P.D.: The Triune Brain in Evolution: Role in Paleocerebral Functions. Plenum, New York (1990) ("Do we owe our intelligence to a predatory past?" C.K. Brain, circa 1923)

Mason, J. (ed.): Exoplanets: Detection, Formation, Properties, Habitability. Springer-Praxis, Heidelberg (2008)

McConnell, B.: Beyond Contact: A Guide to SETI and Communicating with Alien Civilizations. O'Reilly, Sebastopol, CA (2001)

McKay, C.: How to search for life on other worlds. In: Sullivan, W.T., Baross, J. (eds.) Planets and Life: The Emerging Science of Astrobiology, pp. 461–471. Cambridge University Press, New York (2007)

Melott, A., Lieberman, B.: Did a gamma ray burst initiate the late Ordovician mass extinction. NASA Technical Reports Server (NTRS), 6 Aug 2013.

Michaud, M.A.G.: Contact with Alien Civilizations: Our Hopes and Fears About Encountering Extraterrestrials. Springer, New York (2007)

Morris, S.C.: Life's Solution: Inevitable Humans in a Lonely Universe. Oxford University Press, New York (2003)

Muller, T., Koch, E., Wipf, D.: Amino acid transport in plants and transport of neurotransmitters in animals. In: Baluska, F., Mancuso, S., Volkmann, D. (eds.) Communication in Plants: Neuronal Aspects of Plant Life, pp 153–170. Springer, New York (2007)

NAS Colloquium: Neuroimaging of Human Brain Function. National Academy of Sciences Press, Washington, DC (1998)

NASA: Complete Guide to the Kepler Space Telescope Mission and the Search for Habitable Planets and Earth-Like Exoplanets: Planet Detection Strategies, Mission History, and Exoplanets. U.S. Government, NASA, Washington, DC (2013)

von Neumann, J.: The Computer and the Brain. The Sillman Lecture Series. Yale University Press, Princeton, NJ (2012). Foreword by Ray Kurzweil

Nolte, J.: The Human Brain: An Introduction to Its Functional Anatomy. Mosby, Philadelphia, PA (1988)

O'Shea, M.: The Brain: A Very Short Introduction. Oxford University Press, New York (2006)

Plait, P.: Death from the Skies: These Are the Ways the World Will End. . . . Penguin, New York (2008)

Plaxco, K., Gross, M.: Astrobiology: A Brief Introduction. Johns Hopkins University Press, Baltimore, MD (2006)

Ross, M.: The Search for Extraterrestrials: Intercepting Alien Signals. Springer-Praxis, Chichester, UK (2009)

Rothery, D.A.: Volcanoes, Earthquakes, and Tsunamis (Teach Yourself). McGraw Hill, Blacklick, OH (2010)

Sagan, C.: The Dragons of Eden: Speculations on the Evolution of Human Intelligence. Random House (Ballantine), New York (1986)

Sasslov, D.: The Life of Super Earths: How the Hunt for Alien Worlds and Artificial Cells Will Revolutionize Life on Our Planet. Basic, New York (2012)

Scharf, C.A.: Moons of exoplanets: habitats for life. In: Mason, J.W. (ed.) Exoplanets: Detection, Formation, Properties, Habitability, pp. 285–298. Springer, New York (2008)

Schopf, J.W.: Life's Origin: The Beginnings of Biological Evolution. University of California Press, Berkeley, CA (2002)

Schulze-Makuch, D.: Extremophiles on alien worlds: what types organismic adaptations are feasible on other planetary bodies. In: de Vera, J.-P., Sechback, J. (eds.) Habitability of Other Planets and Satellites (Cellular Origins, Life in Extreme Habitats, and Astrobiology), pp 349–365. Springer, New York (2013)

Schulze-Makuch, D., Darling, D.: We Are Not Alone: Why We Have Already Found Extraterrestrial Life. Oneworld, Oxford, UK (2010)

Schulze-Makuch, D., Irwin, L.N.: Life in the Universe: Expectations and Constraints. Springer, New York (2004)

Scientific American (ed.): Evolution: A Scientific American Reader. University of Chicago Press, Chicago, IL (2006). Copyright, Scientific American, Inc

Shostak, S.: Confessions of an Alien Hunter: A Scientist's Search for Extraterrestrial Intelligence. National Geographic, Washington, DC (2009)

Shostak, S., Barnett, A.: Cosmic Company: The Search for Life in the Universe. Cambridge University Press, Cambridge, UK (2003)

Shuch, H.P. (ed.): Searching for Extraterrestrial Intelligence: SETI Past, Present, Future. Springer, Heidelberg (2011)

Skurzynski, G.: Are We Alone? Scientists Search for Life in Space. National Geographic Society, Washington, DC (2008)

Slaughter, M.: Basic Concepts of Neuroscience: A Student's Survival Guide. McGraw-Hill, New York (2002)

Spangenburg, R., Moser, K.: The Life and Death of Stars. Scholastic, New York (2003)

Stevenson, D.S.: Under a Crimson Sun: Prospects for Life in a Red Dwarf System. Springer, New York (2013)

Sullivan III, W.T., Baross, J. (eds.): Planets and Life: The Emerging Science of Astrobiology. Cambridge University Press, New York (2007)

Tarter, J.: Searching for extraterrestrial intelligence. In: Sullivan, W.T., Baross, J. (eds.) Planets and Life: The Emerging Science of Astrobiology, pp. 513–365. Cambridge University Press, New York (2007)

Tarter, J., Impey, C.: If you want to talk to ET, you must first find ET. In: Impey, C., Lunine, J., Funes, J. (eds.) Frontiers in Astrobiology, pp. 175–200. Cambridge University Press, New York (2012)

Thomas, P.J., Hicks, R.D., Chyba, C.F., McKay, C.P. (eds.): Comets and the Origin and Evolution of Life. Springer, Heidelberg (2006)

Tilling, R.L., Helliker, C., Wright, T.: Volcanoes: Past, Present, and Future. Diane Publishing, Collingdale, PA (1987)

Toomey, D.: Weird Life: The Search for Life That Is Very Different from Our Own. W.W. Norton, New York (2013)

Trewavas, A.: The green plant as an intelligent organism. In: Baluska, F., Mancuso, S., Volkmann, D. (eds.) Communication in Plants: Neuronal Aspects of Plant Life, pp 1–12. Springer, New York (2007)

Ulmschneider, P.: Intelligent Life in the Universe: From Common Origins to the Future of Humanity. Springer, New York (2009)

Valoch, D.A. (ed.): Communication with Extraterrestrial Intelligence (CETI). State University of New York Press, Albany, NY (2011)

de Vera, J.-P., Sackbach, J. (eds.): Habitability on Other Planets and Satellites: Cellular Origins, Life in Extreme Habitats, and Astrobiology. Springer, Heidelberg (2013)

Vilenkin, A.A.: Many Worlds in One: The Search for Other Universes. Hill and Wang, New York (2006)

Volkov, A.G. (ed.): Plant Electrophysiology: Signaling and Responding. Springer, Heidelberg (2012)

Walker, G.: Snowball Earth: The Story of the Global Catastrophe That Spawned Life as We Know It. Crown, New York (2003)

Ward, P.: Life as We Do Not Know It: The NASA Search for (and Synthesis of) Alien Life. Penguin, London (2005)

Ward, P.: Under a Green Sky. Smithsonian, Washington, DC (2007)

Ward, P.: The Medea Hypothesis: Is Life on Earth Ultimately Self-Destructive? Princeton University Press, Princeton, NJ (2009)

Ward, P., Bennett, S.: Alien biochemistries. In: Sullivan, W.T., Baross, J. (eds.) Planets and Life: The Emerging Science of Astrobiology, pp 537–542. Cambridge University Press, New York (2007)

Ward, P.D., Brownlee, D.: The Life and Death of Planet Earth: How the New Science of Astrobiology Charts the Ultimate Fate of Our World. Henry Holt, New York (2003)

Ward, P., Brownlee, D.: Rare Earth: Why Complex Life Is Uncommon in the Universe. Copernicus, New York (2004)

Watson, J.D.: Double Helix: A Personal Account of the Discovery of the Structure of DNA. Scribner, New York (1998)

Webb, S.: If the Universe Is Teeming with Aliens.....Where Is Everybody? Fifty Solutions to Fermi's Paradox and the Problem of Extraterrestrial Life. Copernicus, New York (2002)

Whitrow, G.J., Fraser, J.T., Soulsby, M.P.: What Is Time? Cambridge University Press, New York (2004)

Zuckerman, B., Hart, M.H. (eds.): Extraterrestrials: Where Are They? Cambridge University Press, Cambridge, UK (1996)

Index